Openers for Biology Classes

C. Robert R. Gridley

J. Weston Walch, Publisher
Portland, Maine

1 2 3 4 5 6 7 8 9 10

ISBN 0-8251-1682-1

J. Weston Walch, Publisher
P. O. Box 658 • Portland, Maine 04104-0658

Printed in the United States of America

DEDICATION

Openers for Biology Classes *is dedicated to Clifford A. Pulis, director of Camp Beech Cliff for Boys, who made the camping experience an education—an opener for life for his campers and for his staff.*

Not all the great teachers work in classrooms. But all those who touch and shape lives with their example of curiosity, scholarship, and love are great teachers.

Who that well his warke beginneth,

The rather a good ende he winneth.

John Gower

Contents

Biology

Biologists

Birds

Bony Fish

Coelenterates

Crustaceans

Dinosaurs

Echinoderms

Ecology

Environments

Evolution

Ferns

Flatworms

Flowering Plants

Fossil Fuel

Fungi

Genetics

Gymnosperms

Insects

Jawless Fishes

Mammals

Mollusks

Moss

Plankton

Plant Life

Protists

Protochordates

Reptiles

Roundworms

Segmented Worms

Sharks

Soil

Sponges

Symbiosis

Virus

Introduction

"Jacks or better to open," states the dealer.

In poker, a poor hand backed by a good bluff can win—sometimes.

In the classroom, it takes aces to open and no bluffing.

Teachers know the importance of getting students motivated and involved at the beginning of class. *Openers for Biology Classes* is a stacked deck of anecdotes, challenging ideas, stories in miniature, provocative quotations, thought-provoking questions, and interesting facts—all designed to capture and hold student interest in those critical first minutes of class, the time in which students adjudge the day's lesson as boring or worthy of their attention.

How to Use Openers for Biology Classes

The buzzer announced the beginning of the period. It quieted even the last stragglers, who quickly seated themselves.

The teacher read an introductory paragraph to the students, posed a question, and instructed them to "think this over for a minute, then discuss it with a neighbor until I ask you to share your responses with the rest of us."

Stimulated by the introduction and question and pleased to have both an opportunity to think and to share, the students were immediately involved.

The teacher took attendance, passed out a worksheet, then sipped some tea as she watched her students. When she felt the time was right, she called upon several students for answers.

Time elapsed: seven minutes.

Results: attendance taken, housekeeping done, and students already interested and involved with the lesson for the day.

Not bad, for openers.

Openers for Biology Classes was developed to help you get your classes going just that smoothly and effectively—even if you have to take attendance, shuffle papers, or make last-minute preparations after the starting bell.

Openers for Biology Classes has been printed one or two articles to the page for easy use. Each opener has two main parts: a text containing one (occasionally more) activity question to involve students individually or in groups and a set of supplemental activities that provide opportunities to expand, develop, or clarify the theme of the opener.

The *activity question* is printed in red. You may ask the question at the point at which it appears in the text or at any other point you choose.

It is the activity question that hooks the students and provides you with a bit of "free time."

Supplemental activities are printed in a separate section below each opener. These are questions for discussion and library research and invitations to role-play, collect, experiment, debate, or relate biology to other disciplines and relate other disciplines to biology. These questions and suggestions may lead to field trips, independent or group projects, or other learning experiences.

The supplemental activities are phrased as you would present them to your students. You are encouraged to pick and choose supplemental activities to suit your time, resources, and goals. Use them or other activities they bring to mind to stimulate your students to go beyond the lesson, to do and learn more.

Getting Started Using Openers for Biology Classes

Orientation

Browse through the book. Look at the table of contents and index. Read a few openers. Get a feel for the book and its design.

Before Teaching

While you are preparing a lesson, look up the topic and related subtopics in *Openers for Biology Classes*. Look for things, facts, ideas, questions that will help you teach and help your students learn.

You can access the information in this book efficiently using two aids: the table of contents and the index. Each offers special advantages; both take a little explanation—a little time to master.

Table of Contents

The table of contents is organized around 39 general topics: Algae through Viruses. Each entry consists of the title of the opener and a clarifying phrase to give you a general idea of the opener's contents and slant.

This approach is fast, but not especially thorough.

Index

The index is more selective than complete. For example, Rachel Carson is mentioned several times in the book but indexed only once—for Opener 34, *Mornings Strangely Silent,* which focuses on her book *Silent Spring.* Although no opener is devoted to oysters, there are two index entries for oyster, each key to enough information to make your own opener. The point is, if you can find it in the index, you can find enough information to make your own opener! Please note that the index is keyed by page number.

Writing Your Own Openers

Contained within these 200 openers is enough material to write as many more openers. Since an opener is an interesting tidbit made more interesting in the telling, and made relevant by the timing of the telling, writing openers is your stock in trade as a teacher. If this book does not contain the exact opener you want, it may well contain the very information you need to write that opener yourself.

Openers for Biology Classes must go to press incomplete, a resource of motivators just begun. The author, ever the optimist as authors must be, has already opened files for *Openers for Biology Classes II.* Add your suggestions and openers by sending them to me, care of J. Weston Walch, Publisher, P.O. Box 658, Portland, Maine, 04104.

C. Robert R. Gridley

April, 1989

Algae

1. Have You Eaten Your Seaweed Today?

Have you eaten your seaweed today?

Whether or not you realize it, the answer probably is yes. If you brushed your teeth today; if you have eaten any ice cream or mayonnaise, jelly, jam, or cream fillings from junk food; or if you have enjoyed a thick shake at your favorite fast-food place, you have eaten seaweed today.

All of these things and many more are made thicker and creamier by additives extracted from Irish moss and related seaweeds.

Do not feel duped. These additives are pure, nutritious, and low in harmful fats and help preserve the products.

Supplemental Activities

1. Collect edible seaweeds from health food stores, supermarkets, or the seashore and sample seaweed as a food.
2. Peruse ethnic cookbooks for recipes calling for seaweed.
3. Investigate medicinal uses for kelp and other seaweeds.
4. Read labels carefully looking for the words *algin, alginic acid,* or *carrageenan.* Make a list of all products you find with these seaweed ingredients.
5. What is the Irish connection between the words *Irish moss* and *carrageenan*?

2. Oyster Thief Invasion

Unnoticed, the microscopic swarmers established themselves on the oyster's shell. For a period of days, they grew, spread, and turned a bit of chalky white shell chalky green.

The green stain became a cushion of velvety green algae, but still no threat to the sturdy, firmly attached mollusk.

But out of the green mass grew cylindrical bodies that branched twice at their tips. Growing and spreading like the beanstalk in the familiar tale of Jack, the seaweed *Codium* began to threaten the oyster.

How could a seaweed threaten an oyster?

The larger the seaweed grew, the more the oyster felt the tug of the tides and currents on its own body. As the plant made food by photosynthesis, its hollow body filled with gas, making it even more bouyant.

In a rough sea, the persistent twisting and tugging broke the oyster's mooring, setting both plant and animal adrift. On the beach, the oyster would die or feed gulls, another victim of the destructive seaweed known as the oyster thief.

Supplemental Activities

1. *Codium* is just another seaweed on the west coasts of the United States and Europe, but on the American east coast it is a threat to the environment. Explain why organisms that establish themselves in new environments are more often destructive than helpful.
2. Join scientists in speculating how this seaweed might have emigrated to the New World.
3. Biologically, *Codium* is of some interest as a model for the ancestor of early land plants. Look up *Codium* and explain how this might be so.

---

3. Blood in the Snow

As John Ross and his men climbed the cliff that cold day in 1818, they expected a spectacular view of Baffin Bay, but they could not begin to suspect the startling sight that would meet their eyes.

At the top of the cliff, the men gasped, not from the exertion but from surprise—surprise tinted with fear. For eight miles beyond them the snow was stained red—blood red!

What monster or monstrous deed could spread blood over eight miles of snow?

Not if all the Eskimos in Baffinland had been massacred could there have been that much blood. Not if a whaling ship's crew had dragged their catch up the cliff to butcher could there have been that much blood.

Neither monster nor monstrous deed left blood in the snow. The snow was stained red with living algae.

Taking advantage of the sunlight, minerals from seabird droppings, water, and a precious band of warmth at the snow's surface, the microscopic algae were making foodlike plants everywhere—only in a most unusual place.

Supplemental Activities

1. Investigate red tide for the story of algae that poison fish and people.
2. Find out how algae live in hot springs, sloth fur, whale skins, giant clams, and mines far below the earth's surface.
3. Collect algae from rocks, tree bark, soil, or ponds and raise them in the classroom or at home.
4. Learn more about John Ross and his discoveries.

4. The Very Air We Breathe

From your earliest science classes, you learned that plants take in CO_2 and give off O_2, while animals take in O_2 and give off CO_2. Typical of simplifications, this one not only lacks accuracy but also fails even to hint that the plants that give you air are not the familiar ones of farm, field, and forest.

Which plants produce most of the oxygen we breathe?

Two thirds of the oxygen we breathe comes from photosynthesis carried out by phytoplankton: microscopic marine algae.

If you were wondering how important algae could be, consider how important breathing is.

Supplemental Activities

1. Discuss this statement with your classmates and teacher: "What you don't know can't hurt you." Think of the statement in the context of this anecdote about oxygen from marine algae.
2. Research the development of the free oxygen in our atmosphere. The atmosphere of ancient earth carried little of this precious gas upon which all animal life depends.
3. What is inaccurate about the old notion that animals take in oxygen and give off carbon dioxide, while plants take in carbon dioxide and give off oxygen? (Hint: Study the generalized equations for respiration and photosynthesis: $C_6H_{12}6O_2 \rightarrow 6CO_2 + 6H_2O$ and $6CO_2 + 6H_2O \rightarrow C_6H_{12}O_6 + 6O_2$.

---

5. Normandy Invasion

How can starvation be a survival strategy for its victims?

The answer lies on the beaches of Normandy, the scene of a great World War II battle.

The men fighting for the beaches of Normandy on D-Day had neither time for nor interest in the little pale green worms who were also invading those troubled shores.

The *Convoluta* worms hatch during the first really warm days of spring and crawl about the mud feeding on algae.

The algae find their way into the skin of the worms and prosper as the worms feed on the sunny surface mud.

The worms become quite green as they mature. Gradually their mouths seal shut, leaving the worms with only their internal gardens of algae for food.

As winter approaches, the worms exhaust their food supply, becoming pale again. They lay their eggs in the mud just before they starve to death.

Each spring the eggs hatch and little worms reinvade the famous beaches of Normandy. In this way the species survives.

Supplemental Activities

1. Both the worms and the algae benefit from this relationship. Find the technical term for such close relationships between organisms of different species.
2. Research commensalism, mutualism, parasitism, and symbiosis, and be prepared to distinguish each term from the others.
3. Find other symbiotic relationships involving algae and invertebrates.

Amphibians

6. Mad Dog or Sad Dog?

The boys could see their dog coming up the path. The dog staggered, his head lolling from side to side. He stumbled, going down in front. He pushed himself along the ground for several feet before regaining his feet.

Never before having seen their dog behave like this, the boys were puzzled. They watched closely but didn't call out to the dog.

As the dog came nearer, the boys could see frothy saliva covering his snout and drool dribbling from his slack jaw.

"Don't touch Spotty," the older boy cautioned.

The older brother thought he knew what was wrong with the dog. What did he think? What do you think?

"Leave Spotty alone," the older boy ordered. "He has rabies!"

The boys hurried home with the bad news.

Upon thorough examination, the veterinarian confirmed the initial opinion he formed when the family brought in the dog.

"Your dog will be all right unless he tries to eat another toad," the doctor told the anxious family. "Chemicals from the skin glands, the warts, cause those symptoms. Your dog isn't a mad dog, he's just a sad dog—sadder but wiser."

Supplemental Activities

1. The warts of toad skin are glands, not growths. People do not get warts from toads. Where do people get warts from?
2. How poisonous can amphibian skin glands be? This is a question with a surprising answer.

7. The Beginning of the End

For tens of millions of years, amphibians dominated the earth. Their delicate, almost handlike prints were the first real footprints made on the planet. In the steamy swamps of the Carboniferous Period, they waddled and swam about in pursuit of insects, other invertebrates, even smaller amphibians. This time was the Age of Amphibians.

Yet they were not so well adapted for land. Their skin was thin and subject to dangerous drying. Their eggs were, in all probability, jelly-coated—like frog eggs—and their babies fishlike tadpoles. Both to keep their skin moist and to reproduce, the ancient amphibians had to return to water. They were kept close to water, unable to take full advantage of the land environment.

The beginning of the end of the Age of Amphibians was the appearance of a new, super group of vertebrates.

What supervertebrates were to replace amphibians as masters of the land?

Out of the progeny of the amphibians came a hardier stock. The new animals had dry, scaly skin and laid eggs with waterproof shells. These new supervertebrates who were to take over the earth were the reptiles. Their appearance was the beginning of the end of the Age of Amphibians.

Supplemental Activities

1. Trace the amphibian ancestry from lobe-finned fish through reptiles.
2. Justify this statement: After the appearance of reptiles, the world would never be the same again.
3. Amphibia is the smallest, least successful class of vertebrates. Speculate as to why this is so.

---

8. Regenerating Lost Limbs

Among all vertebrates, only amphibians with tails can replace lost limbs and other major structures.

The bones, muscles, nerves, and other tissues of amphibians are not so different from such structures in higher organisms.

Why then can tailed amphibians replace lost limbs when other vertebrates cannot?

Medical researchers find this question very intriguing.

One difference is the amount of electrical activity in amphibian skin. An elaborate sodium transport mechanism generates electricity. Research indicates this electrical activity is a major factor in the ability of amphibians to replace limbs that were amputated and to regenerate surgically removed retinas.

Supplemental Activities

1. Children under 11 years of age have been reported to regenerate perfect finger and toe tips, replacing ones amputated beyond the first joint. Speculate on the significance of this.
2. The edges of wounds to human skin generate electricity. How might this fit into the understanding of limb regeneration?
3. How might knowledge of amphibian limb regeneration be useful to medical researchers concerned with humans?
4. What other animals have regenerative powers?
5. Do some minor organs such as tonsils and appendixes regrow, or are reports of this only folklore?

Arthropods

9. A Fable of Two Kingdoms

Once upon a dream time, two great armored armies confronted each other on the seashore.

Their leaders, great and wise, saw that the impending war would cause much harm and do little good. They determined to settle their differences by diplomacy.

Both sides were seeking dominion over the earth. Division of the earth between the opposing sides would be the only satisfactory conclusion.

It was decided that the land, richer in environments, would go to one group, while the other would rule the sea.

Which invertebrate groups fit the description given in this fable?

The opponents were insects and crustaceans. The insects got the land, the crustaceans the oceans, and only in fresh water would their descendants share common territory.

And such has been the case ever since.

Supplemental Activities

1. After some research into the habitats of insects and crustaceans, comment upon the underlying truth of the fable.
2. Many other biological truths could be explained by fables, folklore, or legend. Compose a fable of your own.
3. A literature teacher or librarian may be able to show you authentic but nonbiological explanations for such matters.

10. Drifters, Swimmers, Plowers, and Burrowers

Like a biological bulldozer, the small animal plowed through the sea-floor sand with little more than its compound eyes showing.

The top of its armored body, horseshoe crablike, was divided into three lobes. Crablike, the little animal trudged on jointed legs.

Others of its kind with tiny, spiny bodies drifted above. Still other species burrowed deeply into the sea floor or swam clumsily in short hops.

For 500 million years these little arthropods would dominate the seas. Their shells would pile up and fossilize, recording eons of life in sequential layers.

What were these drifters, swimmers, plowers, and burrowers?

Trilobites.

Supplemental Activities

1. Find pictures of trilobites and relate their shape to their individual lifestyles.
2. Find and discuss a definition for *index fossil.*
3. Arrange a series of trilobites (fossils, pictures, or models) in order of increasing complexity.
4. Speculate as to how such a large, diverse, and successful group became extinct.

---

11. The Oldest Migration

They were migrating before birds evolved. They were migrating while dinosaurs held the earth. Theirs is the oldest continuous migration known.

Which animals have the longest history of migratory behavior?

Every spring, horseshoe crabs migrate up from their sandy marine shallows to mate on moonlit beaches from Maine to Mexico.

Compelled by instinct, guided by gravity, horsehoe crabs move up-slope, out of the sea and onto the beaches.

The females push through assembled males at the water's edge, literally picking up suitors as they pass. With one to four males riding piggyback, the females push through the sand—out of the waves into the alien world of dry land and air.

The females bulldoze shallow "nests" in sand, lay their eggs—several hundred of them—and then move on. One or more males fertilize the eggs, then follow the females back to the sea.

Mating accomplished, the horseshoe crabs are guided down-slope into the sea by gravity.

Supplemental Activities

1. Many living things make automatic responses to such stimuli as light, gravity, temperature, and chemicals. Find other examples of taxes or instincts.
2. Speculate and then check on definitions for thermotaxis, geotaxis, thigmotaxis, phototaxis, and chemotaxis.
3. Identify some environmental stimuli to which humans make automatic, or nearly automatic, responses. Are these taxes? instincts?

12. Do-It-Yourself Amputation

A New England man amputated his own crushed hand to free himself from a log splitter. His drastic action probably saved him from dying of shock or loss of blood. Among humans such actions are, thankfully, quite rare.

Can you think of some lower animals that amputate their own limbs routinely? How might such self-mutilation be a useful behavior?

One kind of spider throws off legs that have suffered poisonous bites from the spider's mortal enemies: phymatids.

This is a response to poison, a drastic means of preventing the spread of the poison throughout the spider's body.

Supplemental Activities

1. In *The Trees,* one of Conrad Richter's pioneer characters amputated his own thumb with an axe after being bitten by a rattlesnake. Discuss the likely outcome of such first aid. It might be interesting to look up and report on first-aid treatments recommended for snakebite.
2. Starfish, especially brittle and serpent starfish, throw off rays when attacked. How might this be an advantage? Do the rays become "new" starfish?
3. Investigate lizards that are reputed to cast off their tails when threatened. Is this fiction or fact?

---

13. Immortality and the Water Bear

What animal survives death?

For 120 years the dried moss had been in storage. Yet, when dampened by researchers, the tiny water bears, tardigrades, which had once lived upon it lived again!

In 1773 Goeze discovered these tiny eight-legged creatures. Their general shape and lumbering stride reminded him of bears, so he called them *Kleiner Wasser Bär*: little water bears.

Their ability to slow down metabolism for extended periods is more complex than hibernation. The process is called cryptobiosis: hidden life.

How might cryptobiosis be a useful adaption for the water bears?

Adaptations to protect living things from drying out were part of the evolutionary movement of life from the seas to land. The revival of the water bears discussed above is evidence of how effective cryptobiosis is for the water bear.

Supplemental Activities

1. Locate pictures of water bears and consider the similarity the name suggests.
2. Obtain water samples from live mosses and other damp environments and use a microscope to go water bear hunting.
3. Using samples of water bears collected locally or purchased from biological supply houses, dry them out, and see if you can recall water bears to life with a little moisture.
4. Look up the stories of other practitioners of cryptobiosis.

Bacteria

14. Rotting Fish/Starving People

For want of protein, people all over the world starve. The paradox is that many countries in which hunger is a problem have productive fisheries offshore. If only the fish could be delivered, fresh, to the people, they would have the protein they need to live.

How can fish be preserved and shipped to people too poor to bear the costs of refrigeration or canning?

For much of the Third World, where refrigeration is unknown and canning adds too much to the cost of fish, a low-technology solution has been found that brings fresh fish to the people.

Bacteria in the skin of fish are the primary cause of decay. Scientists are getting good results by blanching fresh-caught fish in 194° water, then salting the ice in which the fish are to be shipped. The technology is simple, the cost is low.

Supplemental Activities

1. Explain how blanching and salting contribute to keeping fish fresh.
2. Ice cream was often made at home using salted ice. Why salt ice?
3. Which fish would rot more quickly: bottom feeders or fish from middle water layers? Explain your theory and verify it with research.
4. Japanese researchers have found that a packet of iron oxide, rust, helps preserve canned products. What would rusting do in a closed environment to retard bacterial development?

15. Least, Most, Oldest, Newest

Least, most, oldest, newest.

These words describe bacteria.

Explain how bacteria can be the least, most, oldest, and newest organisms on earth.

Bacteria are the most numerous organisms on the planet. The population of bacteria in a pailful of good humus is larger than the population of humans presently living on earth.

At an average size of a tenth of a micron, bacteria are the smallest, least in size, of all known life forms. That there are 30 trillion bacteria per ounce is testimony to their minuteness.

Fossil bacteria have been found in South African black chert well over 3.2 billion years old.

Recombinant DNA research has produced human-made bacteria, the newest organisms of them all.

So bacteria are the least, most, oldest, and newest of living things.

Supplemental Activities

1. Look into this matter of making new organisms to see what concerns for safety and ethics confront the new technology.
2. An old vaudeville joke concerned a scientist who invented a cure for a disease that hadn't been discovered yet. Write a story of scientists creating a disease organism against which we have no defense.
3. Go beyond the content of this opener to justify this statement: Bacteria are a study in contradictions.

16. Bring Out Your Dead!

Several hundred years ago, in a dusty alley of a small North African seaport, a flea tensed at the approach of a rat. As the rat passed, the flea snapped onto its back, bit the rodent, and hopped off.

On another continent, thousands of humans would die of that flea bite.

The rat, already deathly ill, died in the gutter, its dessicated body its only monument.

Bacteria from the dying rat grew swiftly inside the flea, clogging its digestive system. Ill and near starvation, the flea boarded a passing sailor and bit him, injecting the man with the lethal bacteria.

If the sailor noticed the flea bite, he thought little of it. Vermin were part of his life. Five days out to sea, that bite caused the sailor's death. After a brief but racking bout of fever, chills, and ugly discolored lumps under both arms, the sailor's body was lowered into the sea.

By the hour of the sailor's death, his sickness had spread throughout the crew, the ship's rats, and their fleas. The dead man's ship delivered a cargo of death upon arrival in Italy.

What deadly disease did this ancient ship deliver to Europe?

The bubonic plague would spread so quickly and kill so many, that medieval workers would walk the streets chanting, "Bring out your dead!" in an attempt to keep the pace of burial up to the death rate.

Supplemental Activities

1. The story, although fictional, shows how the plague, or Black Death, spread. Assume you are a Peace Corps worker. Outline steps to protect the people of a Third World country from plague.
2. The nursery rhyme *Ring Around the Rosy* is a relic of the plague epidemics in Europe. Find a complete text of the rhyme and a description of plague symptoms. Then explain the real meaning of the poem.
3. If you decoded *Ring Around the Rosy,* you might be interested to know that most of the other nursery rhymes have hidden meanings, bits and snatches of current events of the time when they were first spoken. With a little help from a librarian or teacher, you will enjoy learning the real meanings of these other nursery rhymes too.
4. Look up the death toll of medieval plague pandemics and compare them with casualty figures for great wars.
5. Examine how the reactions of medieval people to plague compare with modern reactions to AIDS.

Biology

17. The Beautiful, the Mysterious

"The most beautiful thing we can experience is the mysterious. It is the source of all true art and science."

These are the words of a famous scientist, written at the peak of his career.

What do these words tell you about the kind of people who become scientists?

Albert Einstein wrote these words in "What I Believe" for the *Forum* in October 1930.

Einstein was a mathematician and physicist, yet he was to contribute greatly to a project with undeniable biological consequences: the development of the first atomic bomb.

Supplemental Activities

1. Write an essay or poem expanding and developing the quotation.
2. Einstein would later write of the impact of atomic energy and his concern for humankind.

   > The unleashed power of the atom has changed everything save our modes of thinking, and we thus drift toward unparalleled catastrophes.

   Support or attack this statement: Einstein never changed his mind about the beauty of mystery even though he was worried about how we might use science.
3. How might Einstein have responded to this poem by Hermann Hagedorn?

   The bomb that fell on Hiroshima fell on America too.
   It fell on no city, no munitions plants, no docks.
   It erased no church, vaporized no public buildings,
      reduced no man to his atomic elements.
   But it fell, fell.
   It burst. It shook the land.
   God have mercy on our children.
   God have mercy on America.

4. This question has no answer yet, but it is all the more worthy of discussion in light of medical advances, genetic engineering, and myriad other ways biology bodes to affect the human future: What are the moral and social responsibilities of scientists?

18. Word Games

Is it really worth my time to learn all these science words?

Is it really worth it?

If you find yourself thinking that, here is an exercise, a kind of game for you.

Take any word stressed in class or in your textbook. Look up the word in a good dictionary: collegiate or unabridged. Note the derivation of the word, then survey all related words based on that same root.

For example, consider *Aves,* the class name of birds. The word is derived from the Latin word for bird, *aves.* Now consider these related words.

avian:	having to do with birds
aviary:	a bird zoo
aviate:	to fly in an aircraft
aviation:	act or science of flying
aviation medicine:	branch of medicine dealing with effects of flight
aviator:	pilot
aviculture:	raising birds
avidin:	protein in egg whites
avifauna:	bird life of an area
avigation:	aerial navigation
avionics:	science of aircraft electronics

The effort required to add one word to your vocabulary is rewarded, because each new word is a key to more new words and to a powerful vocabulary that enables you to think and speak with precision.

Supplemental Activities

1. Think up a real game based on this opener and play it with your classmates.
2. Comment on this statement: Latin helps people learn English.

19. Nudity and Botany

What do conifers and physical education students have in common?

From the Greek word *gymnos,* meaning "naked," we get our words *gymnosperm* and *gymnasium.*

Physical education students exercise, not in the nude like Greek athletes, but in a gymnasium.

Conifers and other related plants produce seeds that are not protected by an ovary, naked in the sense of being without protection. All such plants are classified together as gymnosperms: the naked seed plants.

The stories of biology words are often as interesting as the things they describe. Any college or unabridged dictionary will give you enough information about the origin of words to piece together their stories.

Supplemental Activities

1. Use a dictionary to find the link between seminaries and sperm.
2. Find a book on word origins in the library for a wider explanation of word stories.
3. Scan the index or glossary of your biology book for words that have spellings similar to words in your general vocabulary. Then check to see if they came from the same roots.

---

20. Speaking Biologically

A startled crayfish flips its tail and abdomen underneath itself and scoots away from danger. From that easily observed behavior comes the slang expression *crawfishing,* backing away from a deal, promise, or obligation.

In how many ways does biology creep into our daily speech?

How many biological expressions can you think up?

Think of people described as: wolves, loan sharks, rats, little lambs, sheep, guinea pigs, goats, asses . . .

We can be sheepish, chicken, mousy, and — well, you get the idea.

Supplemental Activities

1. Are there some biological sayings that might be confusing to people from other states or provinces?
2. The Bible often compares humans to sheep or lambs. Can you explain this particular selection?
3. Make a list of animal names for people that are unflattering or terms of contempt, and a list of complimentary terms or terms of endearment. Compare your two lists. Are there some common characteristics among the animals in both lists?

21. Something About Kicking Cats

Teachers often assign memorization of the major taxa: kingdom, phylum, class, order, family, genus, and species. Such lists of words, especially if the words are unfamiliar and, at least initially, make no sense, can be very difficult to memorize.

What does the sentence "Kick poor cat off fair grounds Saturday" have to do with memorizing the seven taxa?

The sentiments in the sentence are unworthy. The grammar is poor. But the sentence does have two redeeming qualities. It makes enough sense to be memorable, and it will help you remember the taxa.

Compare kingdom, phylum, class, order, family, genus, and species with "Kick poor cat off fair grounds Saturday," and you will see that the words of the list and the words of the sentence start with the same letters. Once you have the sentence memorized, you will have little trouble memorizing the list.

Supplemental Activities

1. What is a mnemonic?
2. Think up mnemonic sentences to help you remember other biological lists and sequences such as mitosis phases, phylum names, and blood flow.
3. Discuss this: Why do mnemonics work, and what do they tell us about how we learn?

---

22. Why on Earth?

In all the vastness of space, from all the scientific research—supermarket tabloid articles not withstanding—we know of only one planet upon which there is life: Earth.

"Why on Earth?" was a popular exclamation of times past. It is also an excellent question to pose when considering the development of life on Earth.

The answer lies with Goldilocks and the three bears.

What could the development of life on Earth have to do with Goldilocks and the three bears?

Earth, like the little bear's porridge, chair, and bed, is *just right.*

Earth is just the right distance from the sun to be within the narrow temperature range favorable to life. Earth is just the right size to hold an atmosphere that is neither too thin nor too thick to support life. Earth's chemical composition has just the right elements in just the right compounds to initiate and sustain life.

Supplemental Activities

1. Find out what is "just right" about carbon, water, oxygen, and other chemicals necessary for life.
2. Read some current scientific views on the possibility of life on other planets.
3. Attack or defend this statement: Just the very number of stars proves there must be life on other planets.

23. What Are the Chances?

The planet, a scant 8,000 miles in diameter, wobbles slightly in its orbit around a mediocre star.

The planet's surface is dominated by sterile seas. The land, occupying less than a quarter of the surface, is barren, geologically active, and volcanic.

The atmosphere, a foul mixture of ammonia, methane, and steamy fog, contains no free oxygen.

Volcanoes spew lava and ash. Lightning flashes from horizon to horizon.

Radiation from the nearby star sears the planet's surface.

What is this planet? What are its chances to develop life?

The chance for life is better than it sounds. This planet is Earth just before life began. The very characteristics that may have made you question the planet's chances to sustain life are believed to be the reasons life began here.

Supplemental Activities

1. Which of the conditions of ancient Earth contributed to life?
2. Look up the experiments of Oparin and Miller and reconcile their findings with this description of Earth before life.
3. Compare the four most common elements now in living things with the formulas of ammonia, methane, and water.
4. If ancient Earth was without free oxygen, speculate as to where our present oxygen supply originated.

---

24. Just the Facts, Ma'am

A famous television police detective of the past generation often demanded, "The facts, just the facts, ma'am."

In biology as in police work, the facts are important.

Compose a statement about the importance of facts in science.

Hear what a famous French scientist had to say of the role of facts in science.

Jules Henri Poincaré said, "Science is built up with facts, as a house is with stones. But a collection of facts is no more a science than a heap of stones is a house."

Supplemental Activities

1. What are the implications of Poincaré's statement for biologists, biology teachers, biology students, and the person on the street?
2. In a letter to Robert Hooke, Sir Isaac Newton said, "If I have seen further [than you or Descartes], it is by standing on the shoulders of giants." Relate this comment to the discussion of facts and their cumulative results.

25. The President Speaks

The president, making his inaugural address, had this to say about science:

> We must embark on a bold new program for making the benefits of our scientific knowledge and industrial progress available for the improvement and growth of underdeveloped areas.
>
> More than half of the people of the world are living in conditions approaching misery. Their food is inadequate. They are victims of disease. Their economic life is primitive and stagnant. Their poverty is a handicap and a threat both to them and to more progressive areas.

Who was this newly elected president of the United States?

The newly elected president was Harry S. Truman!

Supplemental Activities

1. Research the Truman administration and report on how well Truman's inaugural challenge was met.
2. Truman had finished out Franklin Delano Roosevelt's last term and World War II. Report on the scientific achievement that ended World War II and posed a challenge we are still facing.
3. Assume the role of speech writer and science consultant for the next president and write a speech outlining the next administration's prioritize for biology in the United States.

---

26. Words, Words, Words!

Does the vocabulary of biology make you feel as if you are learning a new language as much as you are learning a new science?

Do you sometimes get discouraged with all the terms you must master in your studies?

Consider carefully what Lavoisier had to say about words and biology:

> It is impossible to dissociate language from science or science from language, because every natural science always involves three things the sequence of phenomena on which the science is based; the abstract concepts that call these phenomena to mind; and the words in which the concepts are expressed. To call forth a concept a word is needed; to portray a phenomenon a concept is needed. All three mirror one and the same reality.

Supplemental Activities

1. What were Antoine Lavoisier's credentials as an authority on words and science?
2. Write a simple restatement of Lavoisier's idea.
3. How are scientific observations, concepts, and experiments portrayed other than by words? (Hint: There are several ways!)

27. And Bacterial Creepers

Only in school are there sharp divisions between art, literature, history, music, language arts, and science. Even now you should be seeing that all things studied in school are part of human knowledge, experience, and future.

Hear how well the American poet Theodore Roethke describes biology.

I study the lives on a leaf: the little
Sleepers, numb nudgers in cold dimensions,
Beetles in caves, newts, stone-deaf fishes,
Lice tethered to long limp subterranean weeds,
Squirmers in bogs,
And bacterial creepers.

Why do you suppose so many scientists are poets and poetry readers?

Supplemental Activities

1. Find the full text of *The Minimal* by Roethke to see how aptly he images things biological.
2. Locate other poetic definitions of biology.
3. Can you compare science and poetry? Before answering no, consider what Samuel Taylor Coleridge had to say about that:

> Poetry is not the proper antithesis to prose, but to science. Poetry is opposed to science and prose to metre. The proper and immediate object of science is the acquirement or communication of truth: the proper and immediate object of poetry is the communication of immediate pleasure.
>
> (*Definitions of Poetry,* 1811)

---

28. Questions Not Yet Asked

Are scientists people who know all the answers?

In the introduction of her classic text *Biology,* Helena Curtis traced the major accomplishments of biologists from earliest times. She then ended the discussion by saying, "In our enthusiasm for telling you all that biology has discovered, do not let us convince you that all is known. Many questions are still unanswered. More important, many good questions have not yet been asked. Perhaps you may ask one of them."

Supplemental Activities

1. Paraphrase the main idea of Curtis's statement.
2. After giving the matter some thought, discuss this question: How important are questions to science?
3. Discuss this with your classmates and teacher: What are the implications of Curtis's statement as we join together to answer and ask questions about biology?

29. They Couldn't Print It!

The biologist pulled into an all-night filling station along Interstate 95 deep in rural Maine. After a day of lonely work and several hours of night driving, he was pleased to talk with the youth who was pumping gas.

"If I'd known how dangerous working here was going to be, I'd never have taken this job," the young attendant said as they walked into the garage.

"Holdups?" the biologist inquired, handing the youth his credit card.

"All holdup men do is take the boss's money!" the young man explained. "I'm worried about getting AIDS!"

"AIDS?" The biologist, knowing no connection between pumping gas and AIDS was incredulous.

"Read this." The young man pushed a newspaper of the supermarket-register tabloid variety to the biologist.

"AIDS SPREAD BY MOSQUITO BITES!" was a headline cramped between articles on Elvis's reincarnation as a Siamese cat and a feature story on cannibals in New York City. The biologist skimmed the article while the young attendant machined his credit slip.

"You believe this?" the man asked, struggling to keep a straight face.

"Hey, mister," the youth replied earnestly. "They couldn't print it if it wasn't true!"

How can you tell whether or not to trust things you read?

Supplemental Activities

1. First of all, if there is any doubt in your mind, find out why mosquito bites do not spread AIDS.
2. Make a call to a local newspaper, a friendly lawyer, or other knowledgeable person to see if there are any reasons why a tabloid can't print a story whether or not it is true.
3. After some preparation, have two people continue this story as a skit in which the biologist responds to the young man's conviction that supermarket tabloids are dependable sources of scientific information.
4. Purchase several supermarket tabloids and research a "biological" article to check on the truth and accuracy of the article. This is especially interesting if several people join you, each doing a different article. Then compare your results.

Biologists

30. The Bird Shooter's Apprentice

At 13, Joseph Mason was apprenticed to a master with questionable credentials, a man only recently released from jail for debt. The man was bankrupt, having failed as a farmer, merchant, and miller. The lean woodsman had but two skills: shooting birds and then painting them in startlingly lifelike poses.

Mason and his new master set out south from Cincinnati by flatboat in the autumn of 1820. By Natchez, they were penniless. The master assured the boy that sleeping cold, on muddy ground, with an empty belly would improve the youth's character and art.

In New Orleans, the boy assisted his master painting portraits for food money. Together they wound up on a plantation, where the master tutored the planter's daughter.

Plantation life improved both their arts and their morale. From there the master rose to a fame that young Mason would never approach.

Who was this master hunter and painter of birds?

For the rest of Joseph Mason's life he would speak with pride of his association with John James Audubon.

Supplemental Activities

1. Art and history books will help you learn what young Mason's work and life were like as he traveled with the master bird painter.
2. Read more about Audubon, then write a theme on this premise: It is ironic that the man who may have killed more different kinds of birds than anyone else, before or since, should be a conservation organization's namesake.
3. Look at a collection of Audubon's paintings, full size if possible. Then explain why Audubon, among so many artists of his day, was and still is so highly regarded.
4. From local union leaders, find out how apprentice programs have advanced since Joseph Mason pledged himself to Audubon.
5. If irony amuses you, find out what is ironic about Walt Disney's studio producing the movie *Never Cry Wolf.* (Hint: Remember, Mickey Mouse was Disney's first "star.")

31. Not a Woman's Place

She knew the secret and wrote well of it while most men were only guessing. Yet, because she was a woman, a well-born woman at that, the male scientists of her day did not take her scholarship seriously.

She knew that lichens were not organisms in the usual sense, but algae and fungi bound together in a complex, interdependent relationship.

Her illustrations of fungi, mosses, ferns, birds, and mammals were as accurate and beautiful as any done by male artists. Her writings were as insightful, her scholarship as admirable, as any man's, but such were not "a woman's place."

Instead of science she turned her art and wit to creating and illustrating stories for children.

Can you guess who this talented lady was?

Her books, many of them featuring her most notable character, Peter Rabbit, are enduring treasures left behind by Beatrix Potter.

Yet it is sad that it is only for these stories, not for her intellect and science, that we remember Beatrix Potter.

Supplemental Activities

1. Read books on or by women of science. For starters, try Rachel Carson, Sally Carrighar, and Eugenie Clarke.
2. Sad but true, the work of many brilliant women has been overlooked. Perhaps a librarian or teacher can steer you to other stories like that of Beatrix Potter.
3. Look up other researchers into the mystery of lichens.

---

32. Signature: A Wolf Track

What famous American author, artist, and lecturer signed his works with a wolf track?

With his illustrations and stories, he made the lives and deaths of wild animals important and real to people all over the world.

With his knowledge of woodlore and his leadership, he made the Boy Scout movement a reality in America and influenced its development worldwide.

He was Ernest Thompson Seton, an artist, author, and lecturer whose own life was as interesting and exciting as any of the stories he told or painted.

Supplemental Activities

1. Read and report on a book by Ernest Thompson Seton—you have 42 to choose from.
2. Report on Seton's involvement with the Boy Scouts from his writing of *Woodcraft Indians* through his time as chief scout.
3. After carefully comparing Seton's illustrations with those in Roger Tory Peterson's *A Field Guide to the Birds,* discuss Seton's influence on Peterson.
4. After reading one of Seton's books of animal stories, read a book by Sally Carrighar and compare two authors writing similar stories from different frames of reference. Carrighar's books include: *One Day on Beetle Rock, Icebound Summer,* and *One Day at Teton Marsh.*

33. Beebe of the Depths

William Beebe stared anxiously at the bathysphere as it emerged from the sea. It dangled on a slender steel cable, straining the derrick, obviously heavier than it had been ever before.

Through the thick quartz porthole, Beebe could see that the water that half-filled the diving device was rippling in a most peculiar way. A porthole they had replaced had leaked during this test lowering off Bermuda in 1932.

"Clear the decks!" Beebe ordered before attempting to open the 400-pound door to the diving bell.

What hazard did Dr. Beebe fear as he prepared to open the bathysphere?

Quickly the crew got away from the bell, which as much as anything looked like a giant wrecker's ball hollowed out for passengers.

Carefully Beebe began unbolting the massive door. The pressurized water and air sang, ominously harmonic, as they jetted out. Seconds into the procedure, the huge brass wing bolt securing the door burst from Beebe's hands and hurtled 30 feet across the deck, shearing a half inch of steel out of a pump housing. A blast of water powerful enough to cut a man in half roared out, powered by the pressure that had compacted it in the bathysphere.

Later, the deadly leak repaired, Dr. Beebe would descend half a mile into the inky black water to see fish no human had seen before.

Supplemental Activities

1. Find and read some of the writings of this amazing explorer-scientist. During the 1920's and 1930's he wrote several articles for *National Geographic.*
2. Although the dangers of crushing pressure remain, exploration of the sea depths has come a long way since Beebe. Read about the *Alvin,* the *Trieste,* and other historic deep dives and the people whose courage and ingenuity expand our understanding of the depths.
3. Attack or defend this statement: We know more about the moon than about the depths of the seas.

34. Mornings Strangely Silent

> Over increasingly large areas of the United States, spring now comes unheralded by the return of the birds, and the early mornings are strangely silent where once they were filled with the beauty of bird song.

In her 1962 book *Silent Spring,* the woman who wrote these words stated her belief that we were damaging our environment and would face a fate far worse than springs without bird song unless we altered our ways.

Who was the author and what did she fear was killing the birds and upsetting the entire balance of nature?

Rachel Carson's thesis was that careless use of pesticides was doing great harm to the delicate interrelationships between living things and the environment.

Supplemental Activities

1. Research and report upon the effect of Rachel Carson's book *Silent Spring.*
2. With some help from teachers and librarians, make a list of biology books that have shaped our thinking.
3. Read a book by Rachel Carson: *Silent Spring, The Sea Around Us,* or *The Edge of the Sea.*
4. Rachel Carson is not alone among women science writers. Develop a bibliography of women authors in science.

Birds

35. Mystery of Migration

Why do birds migrate?

By our standards it makes sense: fly south for the winter, fly north for the summer, and enjoy a comfortable climate all year.

Wintering in the south provides more food and a milder climate. Migrating north in spring means breeding in less densely populated territories with an ample food supply.

How sure can we be that this explains migration?

What about birds that migrate in the opposite directions or migrate east/west — or do not migrate at all?

Migration has been studied perhaps more than any other aspect of bird life. For more than a hundred years, humans have studied birds in migration with every tool from simple observations with the unassisted eye to electronic tracking with satellites. Yet we cannot even state with assurance why birds migrate, how they navigate, or how such behavior developed.

Supplemental Activities

1. Some birds migrate by day, some by night. Discuss the advantages and disadvantages of both strategies.
2. Some birds migrate in compact flocks, some in V formations; some fly alone. Discuss the advantages and disadvantages of these behaviors.
3. On a map of North America, plot the major migratory paths, or flyways. Give special attention to your area.
4. Which local birds migrate; which winter in your area?
5. When do birds migrate to and from your area?

36. Mussel Bomber

The falling tide exposed more and more of the Massachusetts shore to the foraging gulls. Among the rock, into the draping seaweeds, the gulls poked for food.

One gull pecked at a mussel but was unable to break the shell. Undaunted, the bird twisted and tugged the mollusk until its anchoring fibers broke. The gull pecked with even less success, for now the free, smooth shell skittered erratically away at every peck.

The gull grasped the mussel in its beak and took off. At about 50 feet above the ledges, the gull dropped the mussel. The impact did what the gull had been unable to do: it broke the shell.

The gull swooped in and gobbled down the rich mussel meat before even the nearest bird on the rocks realized what was happening.

What characteristic, once thought the exclusive property of humans, did this gull demonstrate?

The gull was using a tool to do what it was unable to do with its body. Tool use was once regarded as a significant difference between humans and lower animals.

Supplemental Activities

1. Speculate how such behavior may have developed and how it might be passed through the generations of gulls.
2. Find an old biology book and look up characteristics separating humans from lower animals. How many of these signs of superiority like tool use are now known to be shared with other species?
3. A past president of the Maine Audubon Society called seagulls "flying rats." He intended the comment to be flattering as well as accurate. How are seagulls and rats alike in their adaptation to humans?

---

37. Sweet Firebird

The jack pine barrens of Michigan are home to a bird whose life depends upon forest fires.

How could forest fires be beneficial to birds?

Kirtland's warbler mates only in the jack pine barrens. The pines produce seed cones so dense that only the heat of a forest fire can free the seeds.

Unless fires regularly sweep through the barrens, both the pines and the sweet-singing firebird would be extinguished.

Supplemental Activities

1. What is being done to protect Kirtland's warbler in its winter grounds in the Caribbean?
2. Why are we especially concerned about the future of these trees and birds now? In answering, consider changes in fire detection and suppression and the efforts of Smokey the Bear.
3. Respond to this statement: Protecting jack pines from forest fires is poor conservation.

38. The Adopt-A-Bird Racket

Seven hundred years ago, on a beautiful summer day, a herd of bison grazed lazily. As the massive mammals moved, insects scurried, skittered, jumped, and flew ahead of them. On these insects, flocks of cowbirds fed well and grew plump.

The cowbirds traveled with the meandering herd, flying to pace the herd in stampedes, often riding on the humps and backs of their associates when life was leisurely.

Bison calves trot with mothers moments after they are born, but not so baby birds.

What problems would the usual nesting and baby care of birds present to the herd-following cowbirds?

To keep up with the herds, cowbirds adapted the habit of putting their babies—actually their eggs—up for adoption. The females lay their eggs in the nests of other birds. Cowbird babies, born sooner and larger than nestmates, often dump unborn siblings and hog all the food and care offered by their adoptive parents.

The bison herds would all but disappear with the coming of Europeans. The cowbirds would change their associates to new grazers, horses and cattle, but continue their adopt-a-bird racket.

Supplemental Activities

1. On the basis of the information given, decide whether or not the relationship between the bison and the cowbird is symbiotic.
2. What other birds establish relationships with mammals?
3. Research cuckoos—the birds, not the clocks—for another example of the adopt-a-bird racket.

39. Eat Like a Bird

No expression could be less appropriate for people who eat little or are fussy about food than to say they eat like a bird.

Birds eat a larger portion of their weight daily than even the heartiest eaters in your class.

Why do birds eat so much and gain so little weight?

Birds eat a high-energy diet: seeds, fruit, worms, insects, fish, and meat—maximum energy, minimum bulk.

Their digestion is fast and thorough. A cedar waxwing completely digested a fruit, and disposed of the residue, in 16 minutes flat! The average, however, is a leisurely hour or so for complete digestion in most birds.

The energy from food is expended in flight, in keeping warm, and in simply being a bird.

Supplemental Activities

1. Relate this diet information to flight adaptations.
2. Birds have a high heart rate: 460 beats per minute for a sparrow; over 600 per minute for a hummingbird. How does this relate to flying?
3. Compare bird heart rates with human rates and those of other vertebrates.
4. How apt is the expression "bird-brained" for someone who is not too well endowed mentally?

---

40. Engineered to Fly

"Lower the weight. Increase the power. Maximize control."

These sound like orders for engineers, but they are purposes served by various adaptations of birds to flight.

Can you name some ways birds are adapted for lowered weight, increased power, and maximized control?

Let's consider adaptations that lower weight.

Bird bones are hollow, thin-walled, and reinforced with internal struts resembling the Warren trusses used to strengthen aircraft and bridges. Bird skeletons contain fewer bones than reptile or mammal skeletons: some bones are fused, some simply not present. Bird skulls weigh only a sixth as much as the skull of a mammal of comparable size.

The weight savings are impressive: a frigate bird with a seven-foot wingspan has a skeleton weighing slightly more than four ounces!

Supplemental Activities

1. Investigate ways in which birds adapt for flying by increasing power and maximizing control.
2. Explain how each of these features of bird engineering saves weight: egg-laying, feathers, and lack of teeth.
3. Are bats and insects similar to birds in their adaptations to flight?
4. Respond to this premise: Flying fish and flying squirrels do not fly.

41. Double Breathing

Dissecting a bird reveals structures not found in other vertebrates: a system of air sacs, a sort of biological bagpipe!

How could these sacs help birds fly?

As any athlete will tell you, exercise increases breathing—the body's need for oxygen. Flying demands a large and continuous supply of oxygen. Flying, like other forms of exercise, generates heat. The air sacs address both problems: the need for oxygen and excess heat production.

As birds inhale, air floods their lungs and air sacs. In the lungs, oxygen enters the bloodstream immediately in exchange for carbon dioxide. As birds exhale, fresh air from the air sacs replaces the air leaving the lungs. So, utilizing their air sacs, birds engage in a sort of double breathing.

The air entering the sacs cools the blood, helping the bird discharge the excess heat built up by the strenuous exercise of flying.

Supplemental Activities

1. In dissection or on drawings, locate the air sacs.
2. Here's a challenge: estimate the capacity of the air sacs and compare it with the lung capacity of the same bird. Then compare the combined capacities with the lung capacity of a mammal of comparable size.
3. Using jars, tubing, and balloons, set up a mechanism to simulate breathing in birds.

---

42. Where Does a Half-Ton Bird Sleep?

Where does a thousand-pound bird sleep?

If this were a joke, the punch line would be: "Anywhere it wants to!"

But this is no joke. The answer is: in the prehistory of Madagascar.

Aepyornis maximus became extinct, quite possibly without having been seen by Western people. When its massive bones reached scientists in the West almost 140 years ago, they created something of a sensation.

Aepyornis would have stood nine to ten feet high and would have walked rather than run—most certainly it never flew!

The last of these fabulous birds died some 2,000 years ago, a relic of the burst of adaptive radiation that produced modern birds in the ages following the demise of the dinosaurs.

Supplemental Activities

1. How do *Aepyornis's* dimensions compare with those of Big Bird of Sesame Street fame?
2. Compare reconstructions of *Aepyornis* with those of the terror crane and contrast the lifestyles of the two birds.
3. What is the largest living bird?
4. How would *Aepyornis* compare with the ostrich and emu?

43. The Ornithopter

On a whim, the air force officer pulled into the inventor's yard. The captain parked his station wagon near a strange-looking aircraft: an ornithopter.

The inventor, glad for an attentive, if somewhat skeptical, audience, was soon seated at the controls.

"I have never seen any aircraft fly like this!" the captain exclaimed.

What manner of flight would be unknown to a pilot but perhaps most natural to a bird?

The ornithopter, you see, is supposed to fly by flapping its wings.

The inventor threw the master switch.

Nothing happened.

Several more tries produced only the click of the switch.

When the air force officer, who had tracked down the problem with his eyes, plugged one end of the extension cord into another, the inventor proclaimed, "Ah, I see you are a bit of a technician yourself!"

The ornithopter purred, whirred, flapped, and flopped, but flew not an inch.

Undaunted, the inventor continued his work—so far as the author knows, without success.

Supplemental Activities

1. Discuss this premise: The difference between an inventor and a crackpot is success.
2. *Ornithopter* is a made-up word, based on the Greek word *ornis*. Speculate as to the meaning of *ornis* after considering these words: *Hesperornis, Ichthyornis, ornithology,* and *ornithosis.* If you can get one, you will probably have some idea of what the others mean too.
3. Why would a wing-flapping aircraft be so much harder to make work than the more conventional ones?

Bony Fish

44. Living Fossil

In 1938 a lone African fisherman hauling his net from deep water off the mouth of the Chaloma River, near East London, South Africa, made a discovery that shook the scientific community worldwide.

All the fisherman knew was that what he caught that day was big, blue, and ugly.

What had this fisherman hauled out of the sea?

It was only a skull and some preserved skin by the time it was examined by J.L.B. Smith, but it was enough. Smith knew that he was holding fresh remains of a creature thought to have died with the last of the dinosaurs nearly 70 million years ago!

The fish was a coelacanth, a member of the group called the lobe-finned fishes.

Supplemental Activities

1. Think up a term, more logical than *living fossil,* to describe such finds as the coelacanth.
2. More than 50 coelacanths have been caught in the deep waters off of the Mozambique Channel since the first one was found. Most were caught by fishermen from the Comoro Islands. Locate these places on a map, then read about their geo-history.
3. Find pictures of living coelacanths and their eggs.
4. Run down accounts of other living fossils.

45. Foster Mothering

It was time for the bitterlings to breed. The female was ripe with ova, the male charged with sperm.

After a bit of searching, the female fish found what she sought: a large clam. The fish inserted her ovipositor into the clam's gill chamber and discharged her eggs. Immediately, the male positioned himself over the clam and produced a cloud of milt which was sucked into the clam through its siphon. The baby fish would develop in the soft tissue of the clam.

Half the transaction was complete.

The clam, also a ripe female, discharged its babies, which as much as anything else looked like tiny animated jaws, each little shell having toothlike edges. The tiny clams clamped onto the fins, gill covers, and bellies of the bitterlings. There, eventually protected by cysts of fish flesh, they would develop until ready to live independently.

Both the clam and the fish were foster mothering the other's young!

What advantages would the fish and the clam gain from this arrangement?

Supplemental Activities

1. Speculate how such an odd arrangement might have developed.
2. Consult with your teacher to learn about other cases of foster parenting in nature.

---

46. Caribbean Cannibals

When might cannibalism be a productive survival strategy for an entire species?

Colored to match the canopy of drifting gulfweed, the little sargassum fish were nearly invisible until they snapped at a shrimp.

As they darted about, feeding happily, two green eyes, luminous in the dappled light, ominous in their intensity, followed the feeding fish.

One little fish darted close. A gaping mouth opened. Caught in the inrush of water, the small sargassum fish became the meal of the larger sargassum fish.

The Caribbean cannibal had struck again.

Supplemental Activities

1. Cannibalism, well-documented in these pelagic fish, appears to be a survival strategy. Speculate how eating its own young could help a species survive. (Hint: You might want to read a little more about the fish's environment.)
2. In the last few years, much has been written about cannibalism among humans. Scientists seem polarized between those who see no evidence of cannibalism and those who regard it as a certainty with considerable cultural significance. Read about this debate.

47. The Doctor Is In

The grouper, a massive fish mottled with yellow and brown, sunlight and shadow, had fed all morning on smaller reef fishes.

Now, it hovered menacingly near the coral domain of a tiny blue and black wrasse.

The wrasse was not next on the menu; the grouper was next on the little fish's agenda.

Fearlessly, the little wrasse swam to the predator. The wrasse nibbled fins held erect, mouthed around the giant's eyes and, at last, swam into and out of the gaping mouth.

A tail flicked, and the grouper was away to hunt and feed again.

How do you explain the behavior of these two fish?

Two bearded goatfish swimming in tandem replaced the grouper at the wrasse's station, next in line to be cleaned of parasites.

To these and other fish, the wrasse is a doctor, and to them, the doctor is always in.

Supplemental Activities

1. How might adaptation and evolution explain such a behavior as a small fish ridding others of parasites?
2. Using *symbiosis* as a key word, find other examples of animals that "doctor" others by feeding on their parasites.

---

48. Beef of the Sea

Before refrigeration and canning, preserving food for storage in hot, moist climates was a major concern for explorers, adventurers, and military leaders.

How was food preserved for long expeditions and voyages in those times?

The Grand Banks, a great sandy shoal, supplied one answer to this problem: codfish.

Dried or salted, codfish would last almost indefinitely even under steamy, tropical conditions.

Long before the Pilgrims set foot on a granite boulder on the shore of Plymouth, Massachusetts, Europeans of many nations were fishing the Banks—only a few miles from Canada.

Gradually the British, then the Americans, took control of this valuable fishery.

Codfish stocked almost every ocean voyage during the age of exploration. Codfish fed soldiers, sailors, slaves, and explorers over much of the world.

Supplemental Activities

1. Locate the Grand Banks on a map or, better still, on an ocean chart.
2. Consider this question: How might the skills sailors learned on the Grand Banks have influenced the maritime history of Britain from 1600 through the War of 1812?
3. Research the status, importance, and methods of Grand Banks fisheries today.
4. Sample dried or salt cod to form your own opinion of "beef of the sea."
5. How did cod fishing contribute to the extinction of the great auk?

49. No Resource Is Inexhaustible

"I believe that the cod fishery," Dr. Thomas Huxley stated, "and probably all the other great sea fisheries, are inexhaustible: that is to say, that nothing we do seriously affects the number of fish."

Huxley's remarks were made to an 1883 International Fishery Exhibition. They echoed the belief people had long held: resources of sea, field, and forest were inexhaustible.

How did the technology of Huxley's day lend support to his statement?

Less than a century later, the annual catch of cod from the Grand Banks, perhaps the world's richest fishery, had dropped so drastically that people began to see that no resource is inexhaustible.

Supplemental Activities

1. There were numerous exceptions to Huxley's statement. Can you think of exceptions to the closing statement of this opener? Is any resource truly inexhaustible?
2. Whether or not you agree with the statements above, make a case for conserving a natural resource of your choice.
3. Cite examples of other natural resources that seemed inexhaustible to colonists and early settlers that must now be conserved or lost.

---

50. The Fish Nobody Eats

One of the most commercially valuable fish in the United States is not regularly eaten.

Menhaden, locally known as pogy, porgy, fatback, and bunker, is a giant version of the herring. Menhaden reach a foot or so in length. Even though no one eats them, they are the basis for a major fishing industry.

Why would we fish for a fish no one eats?

The rich, oily meat of menhaden attracts human interest for its useful by-products: oil for paints, lubricants, ink, even soap. The processed flesh, devoid of oils, becomes fertilizer and other products.

Menhaden is the fish that nobody eats but everybody uses.

Supplemental Activities

1. Schools of menhaden numbering thousands of individuals require much food, yet menhaden eat only plankton. Research their unusual way of feeding. It is so very unusual, some fish experts could not believe it until confronted with overwhelming evidence.
2. Are other fish caught only for their by-products? If so, what are they and their by-products?
3. Baby menhaden enter estuaries and develop in their protected waters. Find out why, in our times, this habit may be their undoing.
4. Although people rarely eat menhaden, sharks, tuna, swordfish, and quite a variety of other fish do. How do menhaden survive predation by such powerful enemies?

51. The Great Tilefish Mystery

Never in all his years of fishing had Captain Kirby seen the like of it. His men were hauling in bright blue fish decorated with yellow spots on their rosy to pure white undersides.

They were big fish, averaging a yard in length, firm of flesh and, the ship's cook soon demonstrated, very tasty!

Kirby's crew was fishing just south of the Nantucket Shoals Lightship in more than 150 fathoms (900 feet) of well-fished waters, on a beautiful May day in 1879.

Within two years of their discovery, tilefish, as they came to be called, were the basis for a successful fishery. How such a large population of such large fish had gone so long undetected is something of a mystery.

The greater mystery unfolded in March 1882 with the discovery of acres of dead tilefish floating above their home grounds. More than 1.5 billion tilefish died of causes unknown.

What might have killed this species of fish in such great numbers?

At first, it was thought that the tilefish were extinct, but by 1892, tilefish were being caught again and in increasingly large numbers.

It is now thought that tilefish are very susceptible to water temperature changes and that a minor decrease in the water temperature caused the great tilefish kill of 1882.

Supplemental Activities

1. Since tilefish are back on the menu and in the supermarket, you might want to try this mysterious fish and share your impressions or samples with your classmates.
2. Investigate other reports of mass fish kills.
3. For those who really like to think and research and think again, come up with an explanation as to how such fish might have escaped the notice of the Massachusetts fishermen, men not known for coming back to port with only stories to show for their efforts.

Coelenterates

52. Hydra Everlasting

Abraham Trembley could not believe his own results. In 1744 he cut a section of cells from the animal he had been studying, and from those cells, there grew a whole, perfect new animal.

Trembley repeatedly cut sections from other specimens of this animal, and each section regenerated a whole new specimen.

What was the little freshwater relative of the jellyfish whose immortality so impressed Trembley?

Word of Trembley's discovery caused a sensation, not only among scientists but also among philosophers, literary figures, and the common people. The organism's ability to redevelop from a few cells focused the attention of the world on hydra.

Supplemental Activities

1. How would Trembley's work compare with modern research into cloning?
2. What implications of Trembley's work might have excited so many people?
3. What elements of the story support this statement? Abraham Trembley was a *real* scientist.

53. The Sea Anemone Diet

At the outset of World War II, the researcher joined the legions of young men determined to fight to defend freedom.

The saltwater tanks in which his anemones dwelt were neglected in the years of his absence. The water was seldom replenished, the anemones never fed.

What is there about sea anemones that might have led workers to believe they did not require feeding?

Despite their flowerlike appearance and flower name, anemones are animals, and these animals were not fed during World War II!

When the researcher returned at the war's end to resume his work, his anemones were alive, but greatly reduced in size.

That they had survived the sea anemone diet of no food is testimony to the hardiness of these simple animals.

Supplemental Activities

1. Beginning with starvation and a good encyclopedia, see what you can learn about the ability of other organisms to endure long periods without food.
2. In class discussion, develop this idea: the ability to go without food for extended periods is a survival strategy—to "starve to life."
3. What do sea anemones eat when given the opportunity?

---

54. Sea Wasp

How bad can a jellyfish sting be?

For several weeks after her encounter with a sea wasp, the schoolgirl's wrist was swollen worse than if she had sprained it. The angry weals were stretched open as if they were about to split or tear. It would be a month before the girl would feel really well again.

As serious and as ugly as her injuries were, she was among the lucky ones. Sea wasp stings have proved fatal to people in as little as three minutes. Few poisonous snakes kill so quickly!

The sea wasp is no snake, no insect. It is a jellyfish seldom much larger than your two fists together but more deadly than many larger creatures.

Supplemental Activities

1. Examine prepared microscope slides of stinging cells, nematocysts, to understand their structure.
2. Learn how to treat jellyfish stings and, especially if you live near the coast, learn to recognize any local jellyfish that might sting.
3. Australian lifeguards have taken to wearing pantyhose for protection from jellyfish. How could such flimsy garments protect people from jellyfish? (SAFETY NOTE: The effectiveness of pantyhose has not been established, so you should not count on them for protection—at least not from dangerous jellyfish!)

55. Shipwreckers and Homemakers

Rising from the ocean depths, coral reefs provide, on the one hand, infamous hazards to navigation and, on the other hand, unparalleled opportunities for living things.

Coral reefs constitute the richest, most varied of all ecosystems. They attract, feed, and shelter more different organisms than any other environment. In terms of photosynthesis, their acreage is the world's most productive.

Behind their reefs' massive structure, and underlying their productivity, exists a partnership between creatures from two kingdoms: tiny animals called polyps and tinier plants called algae.

The polyps feed on plankton and deposit cuplike homes of calcium carbonate which form corals. The algae make food by photosynthesis and live within the cells of their hosts, the polyps. Together the polyps and algae solve a basic problem that would prevent either from such spectacular productivity.

What could the algae and coelenterates do together better than either could do separately?

Most of the carbon dioxide the algae need comes from the respiration of the polyps. The excessive carbon dioxide in the polyps raises the animals' pH, triggering rapid deposition of calcium carbonate. Some of the nutrients produced by the plants leaks into the animals.

The polyps provide the algae with a relatively safe, sunny environment.

Together they produce reefs which are both shipwreckers and homemakers.

Supplemental Activities

1. Corals were long considered plants. List reasons for the confusion of the early biologists about corals.
2. Some researchers who advanced our knowledge of coral reefs and whose writings would be interesting to read include Charles Darwin, C. M. Young, and Thomas F. Goreau.
3. Consult your dictionary for the meaning and derivation of the word *atoll.*
4. History and literature teachers can steer you to some interesting stories of shipwrecks on coral reefs.
5. Look up solitary corals and corals of northern waters to learn how some corals manage without forming partnerships with algae.

56. The Lion's Mane

The dying man's last words were "The lion's mane!"

Sherlock Holmes knew from the thin weals that marked the dead man's back that the killer was a strange and monstrous one to so torture and slay a man.

For a while, Holmes suspected the dead man's rival, but the great detective eventually made sense of the victim's last words, then located the real murderer . . .

Who (or what) killed the man in this story?

The detective is most certainly fictitious, but the murderer is very real and very dangerous: a giant jellyfish.

The lion's mane is an oceanic jellyfish of arctic waters. Specimens more than eight feet across the circular body trailing hundred-foot-long tentacles have been observed. The tentacles, armed with thousands of stinging cells, have caused human deaths. Such fatalities are rare only because the jellyfish's range is far north of the places humans swim for pleasure.

Smaller specimens of this tawny to red jellyfish often wash ashore on both sides of the Atlantic.

Supplemental Activities

1. Find pictures of and additional information on this jellyfish in the library.
2. Examine microscope slides of nematocysts, the stinging cells of jellyfish, hydra, and other coelenterates.
3. Look up other dangerous jellyfish: Portugese man-of-war and sea wasp.
4. Find fictional stories involving other dangerous invertebrates—*Conus* snails, for example.
5. Report on personal encounters with jellyfish.

Crustaceans

57. The Ultimate Hand-Me-Down

Cenobita, the hermit crab, scuttled across the scientist's path on the warm sands of a Bermuda beach.

The scientist, A. Hyatt Verrill, took special notice of this land hermit crab because its shell fit so well. Most of the larger hermit crabs Verrill saw were cramped into shells too small to protect their whole bodies.

The well-fitted crab was wearing a fist-sized shell with black and white stripes—the shell of a mollusk that had been extinct for more than a hundred years!

The snails, tasty and easily obtained, had been eaten to extinction by humans. In a very real sense, the hermit crabs had been eaten out of house and home!

"Where, then," Verrill wondered, "did this crab find its home?"

Let's consider that question together.

The answer was not in some hidden sea grotto populated with the last survivors, but in sand dunes of shells long buried. The shell, the ultimate hand-me-down, had last been worn by a hermit crab 120,000 years earlier!

Supplemental Activities

1. Especially if you are unfamiliar with hermit crabs, read about them in a biology book.
2. Hermit crabs, especially land hermit crabs, are easy to maintain in captivity. They are available from pet stores and biological supply houses.

58. More Than Changing Clothes

Do last year's school pants look more like Bermuda shorts this year? Even if the change isn't that dramatic, you have had the experience of outgrowing clothes. Since you grow and your clothes do not, this experience is inevitable.

Lobsters are in somewhat the same predicament. They grow, but their exoskeletons do not. Like all other arthropods, lobsters must shed old exoskeletons and grow new ones.

How would this molting differ from outgrowing clothes?

The process is more complex than changing clothes. The massive claws must be pulled through the narrow joints like Houdini escaping closed manacles. The lobster's shell parts at the joint of the cephalothorax (body) and abdomen (tail). To demonstrate that, you would have to wriggle out of your long underwear through the flap in the back.

The really tricky parts for the lobster are shedding the stomach lining and the covering of its eyes.

Small wonder some lobsters do not survive molting. It is much more than changing clothes.

Supplemental Activities

1. Explain the difference between molting and metamorphosis.
2. Research and compare molting among arthropods representing other classes.
3. Look up these terms used to describe lobsters: shedder, old shell, new shell, and hard shell. Then use the lobster's case to support this statement: You do not always get what you can see.

---

59. Mudbug or Fluss Kreb — They Taste the Same

The Cajun's kettle boils vigorously, wafting the odor of onion, cayenne pepper, lemon, and the nutty scent of crayfish.

These little lobsterlike denizens of fresh water were held in high regard by the chefs of France. So it is a little wonder that they have become part of Cajun cookery and a joy to all people lucky enough to have them.

Call them crawdads, crayfish, *crevlike* (low Dutch) *écrevisee* (French), *camerones* (Spanish) or, by the drawled southern slang or curt German, mudbug or *Fluss Kreb*—they taste the same: delicious.

Can you think of another animal, edible or otherwise, that has a number of different names?

Supplemental Activities

1. Cook up a mess of fresh crayfish by any name of your choice and share them with your classmates and teacher. (This might best be done before dissecting them in lab.)
2. What would you eat if you had to live off the land (if you had only wild things to eat) in your state or province? Many books and fields guides are available if you would like to sample such wild fare.
3. What other commonly dissected biology specimens are edible?

60. Mystery Organ

See if you can figure out what this organ does before you are told.

Lobsters have a small, sand-filled organ at the base of each big antenna. The organs are lined with sensory hairs, richly supplied with nerves, and contain sand grains.

Only when lobsters shed their shells are these organs open to the outside. Replacing the sand grains is part of the shedding process.

Think you know? How could you experiment to test your idea?

Thinking they had an idea, scientists supplied a shedding lobster with iron filings in place of sand. Dutifully, the lobster loaded the organ with the bits of iron.

Whenever a powerful magnet was held over the lobster, the crustacean would flip itself over onto its back. After the magnet was moved away, the lobster would flip itself right side up—no doubt a bit confused.

Are you confused? The organ is called a statocyst, a balance organ which tells the lobster whether it is level, right side up, or otherwise.

Supplemental Activities

1. How does the statocyst work?
2. What organ(s) of the human body are comparable to the lobster's statocyst?
3. Simple experiments are often very effective. For a case in point, find out how physicist Richard P. Feynman demonstrated the effect of cold on the O ring that doomed the Space Shuttle *Challenger*.

---

61. These Fish Are Lousy!

"Fish lice," the biologist explained to her students. She pointed to several flattened, ovalish, nearly transparent animals pressed close to the skin of a Pacific salmon. "They are actually copepods, not insects like real lice."

What would these odd organisms find to eat on fish skin?

Seldom more than a quarter of an inch in diameter, *Argulus*, a typical fish louse, grips its host's skin with a pair of sucker discs and a series of back-pointing spines. Like the insect lice that plague humans and other warm-blooded creatures, these copepod relatives suck blood.

Supplemental Activities

1. Fish lice seldom suck enough blood to be a real problem. Considering that fact, respond to this statement: It is the bite, not the blood loss, that can be the real problem for the fish.
2. If you can collect and examine a quantity of fish, search for fish lice and other ectoparasites.
3. At maturity, the fish lice drop off to breed and then die. Speculate as to how fish catch fish lice, since the adult lice do not breed on the fish.

62. Miracle Glue

This glue bonds to almost anything. It sets quickly, almost immediately, under water. Once set, it resists acids, alkali, protein solvents, and attack by biological agents. It does not crack in extremes of heat or cold. It contains no toxic compounds.

Dentists look at it as a possible boon to their profession. Doctors, especially those who work with bones, hold the possibilities for this glue in high regard. Industrial engineers dream of many applications.

Where can this glue be found?

This glue is not available in stores or through special television offers. It is the substance that bonds barnacles to rocks, shells, and ship bottoms.

Supplemental Activities

1. Justify this statement: Seafaring people all over the world and throughout history have cursed this glue.
2. Barnacles have been studied intensively. Charles Darwin spent so much time on them, his young son, while being shown through a chum's house, asked if the study was "where your father does his barnacles?" Still, synthesis of barnacle glue is beyond us. From the information given in the opener, speculate as to why this glue is so hard to analyze.
3. The giant barnacles of the Pacific west coast are edible. If possible, try some barnacles.

Dinosaurs

63. The Stonesfield Monster

Quarrymen at Stonesfield, England, often found fossils as they split limestone to make roofing tiles. These "imperfections" often caused the rock to break into useless shapes. Although some of the pretty or interesting fossils could be sold as curios, they were more of a nuisance than items of interest or value to the workers.

One rather large jaw had been taken to the museum at Oxford. Several years later, it came to the attention of the Reverend William Buckland.

At his home, the eccentric Buckland lived with a pet bear, Tiglath Pileser, and jackals who sometimes disturbed rectory guests by crunching live guinea pigs behind the sofa. Buckland studied the jaw of the Stonesfield monster.

What was the Stonesfield monster, and what is the significance of its discovery?

In 1824 Buckland presented a paper on the jaw's owner, which he named megalosaurus. This was the first scientific recognition of dinosaurs.

Supplemental Activities

1. Although Buckland's paper preceded the discoveries of Gideon and Mary Ann Mantell by more than a year, the Mantells are often credited as the discoverers of dinosaurs. Theirs is an interesting story too. Check it out.
2. For the story behind the name *dinosaur*, look up Sir Richard Owen.
3. Dinosaurs were widely distributed over North, Central, and South America, Africa, Australia, and parts of Asia, so there may have been important dinosaur finds in your area. Look into this.
4. In all probability, there are dinosaurs yet undiscovered. You might be interested in learning how and where to join the search for new dinosaurs.

64. Monster Birds of Massachusetts

In his mind, Edward Hitchcock saw the giant birds, some of them more than 15 feet tall, striding across the muddy marshes of Massachusetts in some long-ago time.

Unfortunately for Hitchcock, who devoted 13 years to studying the birds and collecting their fossilized tracks, the monster birds of Massachusetts existed only in his imagination.

What animals had made Hitchcock's "bird tracks."

The tracks, which were real enough were actually dinosaur tracks.

For many years, scientists chuckled over poor deluded Hitchcock's bird stories, but recent evidence indicating dinosaurs as the ancestors of birds makes Hitchcock's error seem a little less ridiculous.

Supplemental Activities

1. Make casts of bird tracks and compare them with pictures of dinosaur tracks or real specimens of dinosaur tracks if they are available.
2. Go to the library to find out what part 13-year-old Pliny Moody, a farmboy from South Hadley, Massachusetts, played in the mystery of the monster birds that never were.
3. Indians of many tribes have stories of monsters and giant birds, perhaps based upon dinosaur tracks they saw. Seek out and analyze such stories.

---

65. The Girl Who Loved Dinosaurs

Can an amateur, especially a juvenile amateur, make real contributions to science?

Mary Anning's father was a carpenter with an interesting sideline: selling fossils.

Fossils were desirable curios, collectibles in England 100 years ago. People passing through Lyme Regis stopped at Anning's roadside stand to buy fossils he had collected.

As soon as she was old enough to walk, Mary accompanied her father on his fossil-collecting hikes and soon proved to have a sharper eye for finding specimens than he.

Mary Anning was to discover the first complete skeletons of ichthyosaur and plesiosaur and the first skeleton of a pterodactyl in Britain.

Supplemental Activities

1. For another story of a woman who played a major role in dinosaur discovery, look up Mary Ann Mantell.
2. Dinosaur hunters were often as interesting as the fossils they sought. Look into the story of Frans aron Nopsca, the spy, linguist, and dinosaur hunter, who plotted to seat himself as the emperor of Albania.
3. Several years ago, newspapers reported on a girl whose science fair project on spiders resulted in a major reclassification of arachnids. This might be interesting to follow up.

66. Indiana Jones (?) and the Flaming Cliffs

In the old photograph, the lean, handsome man with the battered felt hat and Colt revolver strapped to his side wasn't the fictional Dr. Jones.

This real-life adventurer-scientist who began his career sweeping floors in a museum was sweeping across Mongolia in search of dinosaurs—not living specimens, but fossils.

Who was this real-life Indiana Jones?

When he stepped away from his dusty truck, planning only to stretch his legs while his guide asked directions from some villagers, Dr. Roy Chapman Andrews made a major discovery. A few feet from the truck, mounted atop a windswept pedestal of sand, was a dinosaur skull. Before Andrews and his party left the escarpment of reddish rocks, he had discovered the nests and eggs of protoceratops, the first dinosaur known to have laid eggs.

Supplemental Activities

1. The adventures of Roy Chapman Andrews are recounted in books that he and others wrote. Go directly to the library for some high adventure reading.
2. Several other dinosaurs have been proven to be egg-layers and nest-builders. Current periodicals have stories of these and other discoveries that are changing our view of dinosaurs.

67. Making Tracks

The young graduate student, with the single-mindedness of a scientist and the boldness of youth, sought dinosaur tracks in the Connecticut River Valley.

Just beyond his reach, he could see a beautiful outcropping of the red shale that often contained dinosaur tracks and other fossils of Mesozoic Era life. All that separated him from the ledge was a fence. It was a high chain-link fence with barbed wire strung above it, but seeing no signs warning against trespassing, the student decided to scale the fence.

Encumbered by his pick and specimen sack, he inched his way up the fence. At the top, doing as John Wayne did crossing barbed wire entanglements in war movies, the student threw his jacket over the wire to pad his crossing. From the top, he dropped into the enclosure.

What situation had the young man gotten himself into in the name of science and adventure?

He was chipping away at the enticing outcropping when he was found by the guards of the minimum security prison he had just broken into!

Supplemental Activities

1. For an exercise in fast thinking, role-play the part of the skeptical guards and the graduate student.
2. Trace the course of the Connecticut River Valley on a map and compare it with maps of Mesozoic formations in a college geology or paleontology text.
3. Are there Mesozoic outcroppings in your area? Unless they are on prison or private property, check them out.
4. Plan a dinosaur-hunting vacation using geology, paleontology, and travel books.

Echinoderms

68. Starfish: Indestructible Enemies

Cursing both passionately and creatively, the oyster fisherman literally tore his enemy limb from limb.

"That'll fix the spiny little devil!" he exclaimed, hurling the starfish's writhing rays overboard.

How was the fisherman wrong in what he did and said?

Starfish destroyed oysters—the fisherman's livelihood—fouled his gear, and made him angrier than an all-day fog. The man had every reason to be upset.

But underwater, only slightly inconvenienced, starfish rays began healing and then regenerating the missing body parts. So long as a bit of the central disc remains, each severed ray can regenerate a complete new starfish.

The fisherman had indeed fixed his enemy. He had created five from the one supplied by nature!

Supplemental Activities

1. In a saltwater aquarium, monitor the fate of detached starfish rays.
2. Seek out other animals and plants capable of regeneration.
3. Speculate as to how studies of regeneration might be important to humans.
4. In local tide pools, or from stocks of dissection specimens, locate starfish showing signs of regeneration.

Ecology

69. Eating Garbage

What happens to the tons of garbage dumped into the sea daily?

One hundred miles off the coast of New York, the ocean floor slopes sharply downward—it is the edge of the continental shelf, gateway to the abyss.

Into the very heart of the ocean, barges are dumping the decaying refuse of the Big Apple and other coastal cities.

It seemed a safe enough place. The mountains of garbage, impressive and distinctive as they are barged by, are minute compared with the vast Atlantic and the great depths off the continental shelf.

But all along the coast fishermen report declining catches, fish kills, crabs and lobsters with shell disease, fish caught but unsalable because of open sores and burnlike tissue damage.

The fish that are salable and the shellfish that show no damage are sold to a nation keenly aware of the health benefits of seafood and hungering for the delicious dishes concocted from sea life.

Are we eating garbage?

Supplemental Activities

1. The question is very important, worth asking again: Are we eating garbage?
2. Role-play the parts of a city garbage disposal officer, a commercial fisherman, a restaurant owner, a supermarket shopper, and a conservation organization representative. Explore the various sides of the garbage disposal problem.
3. Congress was considering anti-dumping legislation as this manuscript was being prepared. Find out how new laws, if any, affect ocean dumping.
4. In a mock legislative session, draft and debate laws to make sure we are not eating garbage.

70. "Let It Burn!"

Ahead of the forest fire, a pine tree explodes into flames. The fire's roar is as loud as it is terrifying. The fire surges forward, a flaming tide, fast on the flow.

The foresters, a university professor, and some Nebraska state officials have a hurried conference.

The officer in charge of the fire-fighting efforts trots to his crew chiefs and their workers when the conference ends.

"Get your crews out of here! We are going to let it burn!" he orders.

Why would experts decide to give up on a fire they might still suppress?

Only a few years ago, the decision to let a fire burn unopposed would have been unthinkable.

Conservationists now recognize that fires, a part of the natural environment, are not always bad.

With no dwellings, no historic sites, no property other than trees imperiled, the experts may decide to let fire recycle the forest. In trees, living and dead, chemicals are locked. The calcium, potash, phosphates, nitrogen, trace elements, and other materials left after fire nourish new growth. Fires keep the life on the land varied. Many animals prosper in the new growth that springs up from the ashes. Sometimes, the best thing to do with a forest fire is to let it burn.

Supplemental Activities

1. A librarian should help you find articles clarifying the notion that fires can become tools for conservationists.
2. Role-play parts of conservationists, favoring and opposing allowing a fire to burn out of control. Add other parts to get a real feel for all the factors underlying such a decision.
3. Ask representatives of your state's or province's conservation department to address your class on forest fires and their use.
4. Draw a cartoon showing someone breaking the news about good fires to Smokey the Bear.
5. Report on the controversy that erupted over the "let it burn" policy during the great fires of 1988 in Yellowstone National Park.

71. Garbage State Beaches

For generations one of the advantages of living in New York City and along the New Jersey shore has been easy access to beautiful bathing beaches. City workers who sweltered all week could loll in cool comfort, enjoy swimming in the ocean, let the kids play in the sand—all only a short drive from home.

Now fist-sized lumps of congealed sewage, plastic bags of decaying hospital wastes, and rotting bodies of fish line the shores. Flags and signs warn bathers to stay out of the water.

Discuss the new meaning of the old saw: "It's OK to go swimming, just don't get wet!"

Supplemental Activities

1. Can you tell some polluted beach stories from personal experience? Hundreds of thousands of people can, and more's the pity.
2. Where does this garbage come from? The question, especially of hospital wastes, is unanswered at the time of this writing, so let us join the experts in finding out.
3. Once we thought the sea could neutralize and absorb all our garbage harmlessly. Justify this statement: We were right and we were wrong!
4. For years New Jersey has been known as the Garden State. How have New Jersey's farmers affected the sea life as they have worked their fields and fed the nation?

---

72. Model of Impending Disaster

A few miles out of Osaka, Japan, the helicopter swoops and turns to follow the course of water stained smoky red. Several other arms of this red tide are visible to the seaward side.

The smoky stains are blooms of algae, tiny plants experiencing population explosions.

The algae may poison shellfish and all the organisms above them in the food chain, including humans. In the shadow of the red tide, sea plants die. The algae will die too, but their decay complicates the problem by depriving the water of oxygen. For want of oxygen, many other species will die.

What has upset the balance of nature in Japan's Inland Sea?

Red tides are natural phenomena, but not on this scale nor in this frequency: more than a hundred a year.

Chemicals run off farm land and dumped into the sea trigger the algal population explosions and their red tides of death.

Japan is a densely populated, industrial land, filled with people who traditionally take much of their food from the sea. Japan's poisoned seas may model impending disaster for the rest of the world.

Supplemental Activities

1. Look up red tide in a botany or biology text and report on the organisms involved and their effects.
2. Discuss this analogy: Food chains are like dominoes in a row.
3. If you live in a coastal area or along the shore of a large lake with a commercial fishery, you may find information on red or brown tides in local newspapers or be able to interview people with first-hand experiences.

73. The Medium Is the Message

Baja California possesses miles of splendid beaches, some of the world's most beautiful and most isolated.

Walk along those beaches as beachcombers always have, and you will find much: plastic milk bottles; plastic bags, with and without garbage; plastic toys; plastic dishes; plastic yokes from six-packs; plastic parts of things no longer recognizable.

Plastic, plastic, plastic.

The medium is the message.

If plastic is the medium, what is the message for us all?

Supplemental Activities

1. What can be done about the plastic plague by individuals, groups, and government?
2. How long does it take plastic to decay?
3. Why is so much of the material washing up on beaches far from "civilization" plastic?
4. What can be done to improve plastic in terms of environmental impact?
5. If you have the stomach for it, find out what happens to animals caught in plastic six-pack yokes or to turtles who mistake plastic bags for their favorite food: jellyfish.
6. How did the search to develop artificial billiard balls lead to the discovery of plastic?

---

74. The Goose, the Grass, and the Slime

In the 1930's, just when many New Englanders were feeling the pinch of the Depression, the wild goose which had graced so many tables for so many years became harder and harder to find.

The wild geese, called brants, fed upon a flowering plant of the sea: eel grass. The eel grass was rapidly disappearing, taking with it a whole ecosystem, including the tasty geese.

Clam diggers and other shellfishers were feeling the loss of the eel grass too as the balance of nature in the estuaries was upset.

What was happening to the eel grass?

The culprit, the killer of eel grass, was, as is so often the case with plants, a funguslike organism: a slime mold.

A series of dry years had increased the salinity of the estuaries, creating conditions optimal for the slime mold—and consequently deadly to the eel grass and the ecosystem built upon it.

Supplemental Activities

1. Research the estuary ecosystem. It is an especially productive one and one very much in danger over much of the Eastern seaboard.
2. Julius Caesar's North African campaign of 46 B.C. owes much of its success to eel grass. Find out more about this interesting connection, thus studying history and biology in combination.

75. Dung and Doom in Australia

With the first settlers came the cattle. For the cattle the land was ideal. For the land, the cattle presented a serious problem: waste disposal.

Why would cattle droppings be any more of a problem than the dung of indigenous animals?

Cow droppings were unappetizing to the local insects—chief among the recyclers of such material. Cow droppings were larger than those of local marsupials. Quickly sun-dried, cow droppings may last for a year or more. A single cow makes enough droppings to cover a tenth of an acre a year. Around the droppings grasses grow lush, but they are unappetizing to cows. The result: unless recycled, that single cow's droppings ruin 20 percent of an acre a year!

How could the droppings of foreign animals be recycled?

The answer was to import dung beetles, insects that eat droppings, recycling them, returning to nature the compounds they contain and cleaning up the pastures.

Supplemental Activities

1. Find out what happened when people brought rabbits and prickly pears to Australia.
2. Look up stories of other alien animals and plants.
3. Are there animals and plants in your area that were imported accidentally or on purpose? Possibilities include English sparrows, water hyacinth, kudzu vines, and the gypsy moth.
4. After reading about alien species and their impact, formulate guidelines for future space settlers of planets upon which life has become established before the settlers' arrival.
5. See Supplemental Activities #1 and #2 for Opener #129.

76. Salt Marsh or Suburb?

Metropolitan areas spread outward along the East Coast of the United States, their boundaries creeping closer and blurring into a megalopolis. People successful in city-based businesses often choose to live in suburbs, in places where they find fresh air, pleasant views, and open spaces.

For these people, some of the most attractive parcels of real estate in the eastern states are salt marshes—broad expanses of grassy land coursed by winding, shallow creeks. Protected from the sea by dunes and beaches, these marshes have a special character of their own.

What other use besides housing could there be for marsh land?

These same marshes are the nurseries for many important species ranging from deep sea fishes to crabs. Destruction of the marshes could forever alter life in the sea—far out into the sea.

Many states must already choose: salt marshes or suburbs.

Supplemental Activities

1. Role-play the parts of land developers, conservationists, local planning officials, and state legislators. Does the interaction reveal how your class would choose between salt marshes or suburbs?
2. Make a field trip—alone or with your class—to a salt marsh and keep a day journal of the life forms you observe.
3. Find current articles that deal with aspects of this quandary and learn the facts and opinions relating to the management and development of salt marsh land.

Environments

77. Forest of Giants

What is the largest of all seaweeds?

Off of California's southern Channel Islands lies an underwater forest of giants: giant kelps.

Macrocysts, the largest of all seaweeds, can reach a length of 150 feet and may grow as much as 2 feet a day! Macrocysts and the slightly smaller elk kelp form dense, tangled forests in these clear, cold waters.

In a fluid world of luminous light and mottling shadows, a very special ecosystem functions. Amidst the giants of the algae world, phytoplankton pastures are grazed by microscopic herbivores, hunted by tiny predators. These little lives and the kelp itself nourish complex and interacting food chains including sea urchins, abalone, sea hares, and octopuses, and culminating with seals and sea lions and occasionally the ultimate predator: the great white shark.

A single kelp plant may support 500,000 organisms — a world unto itself.

Supplemental Activities

1. Visit this underwater ecosystem even if only through magazine articles.
2. After studying the ecosystem, write a short story featuring human explorers or some indigenous species in these forests of giants.
3. Research uses humans have found for kelp.
4. Study kelp harvesting and farming.

78. A World Adrift

What fearsome body of water would leave sailors adrift without a breeze to move their ship?

On a map of ocean currents, the Sargasso Sea appears to be a giant eddy in the Gulf Stream two thirds the size of the United States. It is visually defined and named for sargassum, the seaweed that dominates this unusual ecosystem.

Above feeding fish, the roof of their world glows golden and shades to murky brown as sunlight illuminates the floating sargassum.

Crabs, snails, and worms that never in their lifetimes will know solid ground forage on floating seaweed—all adapted to the pelagic lifestyle.

The seaweed, with its bubblelike floats and golden brown leaflike blades, gave up life along this seashore millions of years ago. It broke away from the land carrying like the ark of Noah its population of shore creatures far out to sea, there to evolve into a unique world adrift.

Supplemental Activities

1. Locate the Sargasso Sea on a map of the Atlantic Ocean.
2. Read accounts of sailors becalmed in the Sargasso Sea.
3. Look up animals and plants of this ecosystem and give particular attention to their adaptations: changes that camouflage, protect, and empower them to deal with life adrift.
4. Speculate as to the advantages of gas-filled bladders, floats on seaweeds that live anchored to the bottom.

79. Sloggers

In a single file, chest high in grass, wading up to their knees—sometimes to their hips—in slow-moving water, six young Americans marched carefully. At a hand signal from the youth on point, the whole group froze in their tracks. Only their anxious eyes moved, straining to see what the leader had seen.

The scene looked like clips from films of the war in Southeast Asia. The young people were high school students, and their marsh was in Montana.

Why would half a dozen reasonably intelligent people slog through a marsh?

Their mission was to seek and discover, not search and destroy.

In the marshes live animals and plants specialized to a rich world of water, mud, and each other. Frogs snap at insects, bloodsuckers seek their prey with sinuous swimming, water scorpions—dangerous only to fellow insects—wait in ambush, and carnivorous plants grow in abundance.

The world of the marsh is a world of adventures waiting to be experienced by sloggers.

Supplemental Activities

1. Return from a slog with enough specimens to start an aquarium so others, not so daring, can see a bit of the marsh and so studies can continue.
2. For the really ambitious, consider establishing a farm pond on the school campus. In schools with even moderate grounds, small ponds can be made. A marshy pond can be an asset to a school in many ways and a good project for a biology club or other school-spirited group.

---

80. Under White Water

Is there life under white water stretches in rivers, or is the bottom swept clear of life—a biological desert?

Out of the hills and mountains of North America, countless brooks, cold and clear, cascade seaward or to their termini in lakes. Fast-moving water, sometimes humping up white and frothy, sometimes running with treacherous slickness over rocks, might not seem a hospitable environment for living things, but an amazing number and variety of organisms—representatives of every major phyla—have adapted to life under white water.

Even in the swiftest currents, algae mat the rocks, pastures of "grass" to be harvested. Larval insects such as water pennies, caddis fly nymphs, and others find both footing and food on the slimy rocks. In the inevitable eddies and sheltered nooks, fish rest and feed. The mud between and under rocks supports worms, especially roundworms, and a world of detritus feeders.

In the swift waters there are algae scrapers, bottom burrowers, filter feeders, and swift predators like salmon and trout.

Supplemental Activities

1. Using a field guide to freshwater life for your area, study the kinds of animals and plants that live under white water.

Evolution

81. Words That Shook the World

> The Simiadae then branched off into two great stems, the New World monkeys, and the Old World monkeys; and from the latter at a remote period, Man, the wonder and glory of the universe, proceeded.

Who was this author who dared to suggest that humans descended from apes?

The notion that humans might have evolved from anything, let alone Old World monkeys, shook the world when Charles Darwin uttered the idea in 1871, and it has continued to shake the world for more than a hundred years since.

Supplemental Activities

1. View the movie *Inherit the Wind*, then discuss the reaction of Americans to the theories of Charles Darwin.
2. Find writings describing the Scopes trial, upon which *Inherit the Wind* is based, and compare fiction and fact.
3. With the help of a librarian or a representative of your local newspaper, locate the most recent articles about teaching evolution and discuss the various sides of this issue.
4. Look up Wilberforce and Huxley to see how learned men of Darwin's day reacted to his theory of human evolution.
5. Ask your parents or religious leaders how they view the theory of evolution.
6. Secure several different high school biology texts and compare their presentations of evolution — especially human evolution.
7. Read "The Case of the Creeping Fox Terrier Clone" in the This View of Life section of the January 1988 issue of *Natural History* as a supplement to your work on Supplemental Activity #6.

82. 10^8 Years in 24 Hours

What is the purpose of making scale models?

Here is a scale model of geologic time. The scale is 10^8 years = 24 hours. All the evolution of life is scaled down to a size that can be meaningful.

Imagine it as the ultimate movie, a film spectacular, a 24-hour presentation.

At the stroke of midnight, even as the curtains part, there is a flicker of activity on the screen: chemicals have become the first living things! We watch molecules eating other molecules, gradually becoming single-celled animals and, sometime after sunrise, single-celled plants.

At 4:30 P.M. seaweeds and invertebrates of many forms virtually explode into view. At 6:45, the first plants invade the land.

By 7:50, lobe-finned fish flounder onto land as the first amphibians and pursue the insects, which preceded them only by seconds.

A little after 9:00, the first reptiles evolve. At 9:30, a chicken-sized, two-legged, reptilelike creature struts into view and for the next hour and a half, the dinosaurs dominate the earth.

At 11:00 P.M. the screen darkens mysteriously for an instant before illuminating a world suddenly without dinosaurs. Birds and mammals vie for supremacy.

Two seconds before midnight, the first humans appear. All written history flickers by us in the last quarter second before midnight.

Supplemental Activities

1. With a good geology book, an inexpensive calculator, and some spare time, make up some scale models of time; perhaps you could expand the story of life to a month or a year, or condense it to a school day.
2. *Model* is a word with special meanings in science. Find out what kinds of models scientists use as they try to understand the world around them.

83. Smarter Than the Devil

How would people explain fossils if they had no knowledge of science?

The monk's hands trembled as he picked up the shell of stone. You would recognize it as a fossil, the petrified remains of one small seashell, its original minerals replaced by others until it is truly a shell of stone. You might marvel at it, but your hand would not tremble.

The monk of medieval Europe, who truly believed that God had just discarded this model for a life that might have been, trembled to have something in his hand that had just left the hand of God.

Other medieval scholars considered fossils to be proof of the great flood through which Noah piloted the ark and its cargo of lives spared.

Still others prided themselves in recognizing fossils as foiled attempts of the devil to trick them. For if Satan tricked them into believing a stone to be a shell, he might next trick them out of their souls. How comforting to know you are smarter than the devil!

Supplemental Activities

1. Prehistoric graves often contain fossils deliberately placed there by ancient humans. Speculate as to what beliefs might have prompted the survivors to bury fossils with their dead.
2. Shamans often number fossils among their accoutrements to lay and remove curses. What significance might they see in shells of stone?
3. Do any modern religions hold fossils in special regard?

---

84. Return to Rhynie

Where would you go to search for fossils of the first land plants?

In the 1920's two botanists went to Rhynie on the Scottish coast to see what more could be learned about the earliest land plants.

In fossil-bearing strata a generation earlier, a medical doctor had stumbled across the earliest known vascular plant: *Psilophyton.*

These two botanists would discover no earlier plant, but they would discover three other links in the story of land plant evolution: *Hornea, Asteroxylon* and, most importantly, *Rhynia,* whose fossils included spore cases.

Supplemental Activities

1. Find Rhynie on a map of Scotland and on geologic maps showing the ages of exposed strata.
2. Make a family tree of land plants beginning with the hypothetical ancestral seaweed and continuing through the spread of flowering plants.

85. The Ancestor

How would seaweeds have changed to become the ancestors of land plants?

Five hundred million years ago, the Ancestor[1] lived along the shores of warm, shallow seas. The Ancestor could endure the low-tide times of dryness. It had a peculiar, at least by human standards, way of reproducing. One generation would reproduce by sperm and ova, sexually; the next generation would reproduce by spores, asexually.

The sexual generation depended upon water for the sperm to swim to the ova, but the asexual generation could spread spores by the wind.

Gradually, the Ancestor's descendants adapted themselves to land and the new opportunities land provided.

This Ancestor was the ancestor of all land plants. Of the Ancestor, we have no fossils. It is a mental model of an organism that took a major step in evolution: pioneering life on land—opening the way for all the plants and animals to follow.

Supplemental Activities

1. Research *Rhynia,* seed ferns, cycads, ginkgoes, and other important plant fossils.
2. Examine specimens of *Codium* (available from biological supply houses) or other primitive seaweeds.
3. Using a geologic time scale, find the name of the period during which the Ancestor lived and see what else was happening then.

---

[1] Note: The term *Ancestor* is a word of convenience, not a term commonly used by any other author.

86. The Laid-Bare Plants of Scotland

What kind of people find the important fossils that shape our view of life in the past?

In a great many cases, untrained amateurs have been the discoverers of important fossils. Consider this as a case in point.

Dr. Mackie found some unusual fossils near the Scottish village of Rhynie early in the 1800's.

Recognizing their uniqueness, he sent them to Sir William Dawson, a famous botanist of that time.

Dawson would name the plants *Psilophyton,* the laid-bare plants, and describe their place in the evolution of plants.

In the Ordovician Period, when that part of Scotland was a muddy seashore, the stems of *Psilophyton* spread out, sending up upright branches at intervals. The branches, totally leafless, spread out by forking.

Psilophyton, the earliest known plant with conductive tissue, was discovered by a doctor who enjoyed nature and strolling along the seashore.

Supplemental Activities

1. Sketch your impression of *Psilophyton* from this description, then seek out illustrations to check your work against.
2. React to this statement: Of the roots, stems, leaves, and flowers common to land plants, stems came first.
3. Obtain specimens of *Psilotum nudum,* the whisk fern of Florida and Bermuda, for living examples of plants that have changed little since the shores of Rhynie began to grow green.

87. The First Seeds

What were the first seed plants?

Gilboa, New York, is a small town southwest of Albany. Even today it has less than 1,200 citizens, yet among scientists it is world famous.

Gilboa is the site of a discovery of great importance in our understanding of plant evolution.

About 100 years ago, surging flood waters exposed a petrified forest at Gilboa. Some of the trunks were ten meters high, more than a meter wide at their bases.

These plants were not trees. They were the remains of fernlike plants.

During the Devonian Period, the Age of Fishes, these graceful, tree-sized ferns grew in dense forests.

What made these fossils especially important?

A team of British researchers made a startling discovery as they studied these fossils: these ancient plants reproduced by seeds! Today we regard them as the first seed plants.

Supplemental Activities

1. Study the life cycle of modern ferns.
2. Look up cycads, ginkgoes, and *Rhynia*—other important links in the chain of events leading to modern plants.
3. Contact the museum of Gilboa for further information.

---

88. Superplants!

Near the end of the Mesozoic Era, the Age of Dinosaurs, one group of gymnosperms became the superplants.

They changed their way of reproducing, became very successful, and boosted the evolution of insects, birds, and mammals—perhaps even hastening the development of humans!

What are these superplants?

They began producing seeds within ovaries which developed from flowers. Their particular manner of reproduction was ideally suited for land. Their seeds, packed with energy, provided developing birds and mammals with concentrated, high-energy food. As these plants spread over the land, they took insects, birds, and mammals with them. The superplants are the flowering plants!

Supplemental Activities

1. What is there about cycads that suggests they are the ancestors of flowering plants?
2. Respond to this premise: Without flowering plants, humans would have little to eat.
3. Justify calling flowering plants superplants.

Ferns

89. Of Ferns, Spies, and Spying

How useful would it be to be able to become invisible?

Ferns have played quite a part in the folklore of Europe.

The evergreen Christmas fern has long been regarded as a symbol of the life everlasting promised by Christianity.

Oak fern root stocks have a partial circle of conductive tissue which in cross section looks like a letter *C*—the initial of Christ. The old belief was that the more perfect the *C*, the better the luck the fern "root" would bring!

Perhaps the most charming belief concerning ferns was the magical use of fern dust (spores). According to folklore, whoever sprinkled fern dust upon her or his shoes on Saint John's Eve would walk all the next day totally invisible!

Supplemental Activities

1. Excepting fiddleheads, undeveloped fronds, ferns are not used for food or medicine. How does it happen that such a store of folklore built up around plants so little used?
2. Fiddleheads, pickled and canned as well as fresh, are frequently available for sampling and class sharing.
3. Besides ferns, what other plants have evergreen members?

Flatworms

90. Remarkable, Remarkable Worms

Which invertebrate can have two heads, two tails, even two bodies?

The answer is planaria, the remarkable, remarkable worm. Regeneration experiments show planaria capable of growing new heads, tails, and bodies. Split heads or tails, held apart while they heal, produce worms with double heads, double tails, or both.

A scientist trained planaria to go through a maze to get food. Planaria who had learned the maze were ground up and fed to "untrained" planaria. The cannibals ran the maze with ease, apparently having learned from lunch. Are smart pills on the horizon?

Supplemental Activities

1. Collect live planaria from local streams, or buy them from biological supply houses, and raise them in the laboratory.
2. Conduct your own regeneration experiments. You can find directions in laboratory manuals for high school and college texts.

91. Liver Rot

In 1793 Jean de Brie discovered the cause of a puzzling disease of sheep by dissecting one victim. When he opened the liver, wriggling flatworms spilled out onto his table.

They left de Brie with more questions than answers.

How did worms get into the sheep? What could be done to protect sheep from the worms?

For the next 154 years scientists, including Anton van Leeuwenhoek, gathered bits and pieces of the mysterious life cycle of the sheep liver fluke.

It was not until 1893 that Adolf Lutz, a Brazilian scientist, filled in the last details of this parasite's life.

The parasite develops through stages, each with a distinctly different body, suited to a different environment or host.

The eggs hatch in sheep feces; successive stages swim about, attack certain snails, gnaw out, encyst themselves on water plants, and complete their life cycle in the bodies of grazing sheep or hapless humans.

Supplemental Activities

1. Examine specimens available from biological supply houses in situ and free.
2. Report in detail on the life cycle of the sheep liver fluke.
3. OBSERVING ALL SAFETY PRECAUTIONS, examine sheep feces or those of other domestic animals or pets to see if parasites or their eggs are present.

92. Swimmer's Itch

"I'm all itchy," complained a little girl. She and her older brother were splashing ashore in a Pennsylvania pond.

Their mother toweled the little girl down with a big beach towel, but the boy scooted off to buy a cone of soft ice cream and let the sun do his drying.

On their way back to Harrisburg, both children showed symptoms of their encounter with swimmer's itch: raised spots on their skin as much as a quarter of an inch in diameter. The boy's skin had more and larger blotches, and he was running a low fever by evening.

During the next two weeks, both children recovered without further complications.

What had happened to the children to cause them this distress called swimmer's itch?

Swimmer's itch is caused by parasitic flatworms. The worms had mistaken these children for their usual hosts: water birds.

The parasites mature and breed in the intestines of water birds. Their eggs leave with the hosts' feces to ripen and hatch in fresh water. Free-swimming larvae are released from the eggs to seek an alternate host: snails. Three body changes later, the parasites exit the snails and swim frantically about seeking water fowl. Occasionally, as with the children, the parasites encounter human skin and burrow in, causing swimmer's itch.

Supplemental Activities

1. How had the mother's behavior lessened the reaction of the little girl?
2. Explain this statement: The parasite's error is fatal.
3. Find out if swimmer's itch is ever a problem locally. If it is, find out which organism(s) are involved and work out a chart showing their life history.

93. Gut-Busted Fish

The trout lay limp in the biologist's hand, its intestines protruding obscenely from its gill cover.

What would cause a fish's intestines to hand out of its gill slits?

"Here's another one," the researcher said to the graduate student engaged in hauling in more of their seine.

"Lingula?"

The biologist nodded.

Nearly three years earlier this trout had sealed its fate by eating a copepod—a bit of freshwater plankton.

A heron infested with the tapeworm *Lingula* had left some feces filled with parasite eggs in the water. After six days, the eggs had released swimming larvae. The larvae were eaten by the copepod, the copepod by the trout.

The form infesting the trout had grown so much it displaced the unfortunate fish's intestines. The trout did not die alone. Having destroyed its environment, albeit accidentally, the parasite had destroyed itself as well.

Supplemental Activities

1. How would the parasite have gotten to a water bird such as a heron?
2. After some thought, discuss how the fate of the parasitic worm *Ligula* parallels the human condition in our polluted world.
3. After considering these figures, expand upon this statement: The parasitic lifestyle is a hazardous one.

Stage of Tapeworm	Development Time Required	Time to Find Next Host
Eggs	5–8 days	2–3 days
Copepod Stage	9–10 days	3–5 days
Fish Stage	1–2 years	within fish stage
Bird (mature tapeworm)		

4. A common strategy in attacking a parasite is to eliminate a stage in its life cycle. For example, we attack malaria by killing the mosquitoes that carry it. Which stage of *Lingula*'s life cycle could be interrupted without causing unacceptable environmental impact?
5. Can you relate the old saying "There's no such thing as a free lunch" to this story and these questions?

Flowering Plants

94. Staff of Life

How have we come to call money dough? How have we come to call the family money-maker the breadwinner?

In our economy, money is wealth. But not long ago a person with a smidgen of cheese, perhaps some herbs, and a slab of bread daily would have been prosperous.

Except in Asia, where rice fills bellies and bread was not made, bread has constituted half or more of the diet of the common people.

Bread made of wheat, oats, rye, or other grains was then and is now a tasty, nutritious, and convenient food. Loaves were fashioned into plates, portions, even decorative shapes. Loaves, whether homemade or bought with the coins of the realm, were the culture's product, the baker's pride, and the possessor's joy and meal.

Small wonder, then, that *bread* and *dough* became synonymous with money. Bread was the very staff of life.

Supplemental Activities

1. Consult foreign language teachers to see if there are parallels for *bread, dough,* and *breadwinner* in the idioms and slang of other languages.
2. Do other words in our language equate wealth with food?
3. Find out what shapes loaves have had in different cultures and different times.
4. For a real project, make portion bread, trenchers, bowl bread, crossbread, or other traditional loaves for the class to see and sample.

95. Initiation Insanity

In times before either tribe ever saw people of white skin, boys in Algonquin villages and boys in Amazon River tribes became men in rituals involving powerful hallucinogenic drugs.

In South America the initiates drank caapi to see and dance with tree spirits. In the cooler forests of the Algonquins the braves-to-be drank wysoccan, a jimsonweed brew, then hallucinated ritual death and rebirth as men.

What's the difference between this use of drugs and the kind of drug use most parents now condemn?

The differences are many and worthy of consideration, especially when we occasionally encounter the argument that drugs are "natural" and "have been used for hundreds of years."

In these tribes and many others, especially in South America, the use of the drug is ritual, not recreational; supervised by adults, not done on the sly; once in a lifetime, not habitual; and legal.

Supplemental Activities

1. Research the uses of scopolamine.
2. Find out what happens to cattle and other livestock that graze on jimsonweed.
3. After reading about both, compare primitive initiation rites with college fraternity and sorority initiations.

96. Portable Pastures

What makes weeds different from other plants?

The captain of the Philippine fishing boat swore in eloquent and sincere Spanish and Filipino. The plants that fouled his propeller would cost him the tide and his head start. To him the water hyacinths were weeds — harbor-clogging, equipment-fouling nuisances.

These rootless plants grow in circular clusters, some as large as suburban lawns. Almost without natural enemies, water hyacinths spread rapidly. They obscure navigational buoys, reefs, and channels; they foul nets, propellers, pumps, and anchor cables.

However, these circular clusters could be towed from places where they are nuisances to places where they would be welcome producers of fodder for livestock. They might be used to clear polluted harbors because they take in and recycle many common pollutants.

Is hyacinth a weed or a crop?

To the Philippine captain, to many people, water hyacinth is a weed, a useless, troublesome plant. To waterfront dwellers and farmers of the not-too-distant future, the water hyacinth may become a useful crop, a crop that grows in portable pastures.

Supplemental Activities

1. Kudzu vines and prickly pear cactus are also prolific breeders which soon overrun areas. Their stories make interesting reading.
2. Speculate as to the uses of a fast-growing, rootless water plant in the future, in a hungry world, in space.

---

97. Tonic, Treat, or Fodder

What vegetable from Afghanistan became a favorite of American Indians?

On the windswept highlands of Afghanistan, the lacy-leafed, tap-rooted plants grew first.

The Greeks regarded the plant as a digestive tonic to be eaten when they had eaten too much of other things.

The Romans disdained this plant as food, but maintained it improved the flavor of lamb if used to fatten the animals before slaughter.

As the vegetable spread across Europe, people began to eat it as we do today. It became a staple which settlers took with them to the New World.

The Jamestown colonists planted and shared these tasty roots with their Indian neighbors, who found them such a treat they traded them with other tribes. So widely were the seeds traded that Lewis and Clark concluded the plants must have been indigenous when they found Plains Indians eating carrots.

Supplemental Activities

1. Tracing the origins of vegetables is of more than historic interest. From ancestral stocks, new, improved species might be developed. Try to find out where your favorite vegetables first grew.
2. Do carrots have medicinal properties?

98. Asia's Fruit of Gold

Where did the orange in your lunchbox come from — originally?

The first oranges grew in the tropical forests of southern China, possibly in Vietnam or Cambodia.

The details of their spread to Spain and Portugal are obscure, but we know that orange seeds came to the New World with Christopher Columbus.

By the 1500s, oranges had been established in Central America, then in Mexico.

The settlers of Saint Augustine planted the first Florida oranges.

Despite their long history of cultivation in the New World, oranges were a luxury item in the United States as recently as the turn of this century.

Grandparents may remember oranges as special treats in Christmas stockings rather than as daily lunchbox contents.

Supplemental Activities

1. Interview your grandparents or other elderly people who might remember oranges as a treat in their childhood days. Ask them about other changes in diet they have seen in their lifetime.
2. Work out a family tree of oranges: sweet, bitter, and mandarin.
3. Trace the origins of other fruits and vegetables.
4. For a really challenging project, prepare a world map showing the origin and spread of fruits and vegetables.

---

99. Snakeroot

Out of the tall grass, without warning, the cobra struck. The boy screamed. His parents, peasants in ancient India, sped to this side. Moments later, the boy's little sister was racing to the healer to get a bit of dried, shriveled root to treat her brother's bite.

What are the chances of survival for this boy of India treated with snakeroot?

In 1,000 B.C. merchants in the dusty bazaars were offering *Rauwolfia* roots as a cure for snakebite — something of a problem in India, even today. It is sold there, just that way, still.

From that folk medicine use, the plant takes its common name: snakeroot.

Snakeroot, *Rauwolfia serpentina,* first described in western scientific literature by Leonhard Rauwolf in the late 1500's, is still held in high regard as a medicinal plant.

Derivatives of snakeroot are used as sedatives and most importantly as a drug for combating high blood pressure.

Supplemental Activities

1. Translate and then explain the meaning of snakeroot's technical name.
2. Suppose you were the discoverer of snakeroot. What would you name this medicinal plant?
3. Some research might uncover the value of snakeroot as a snakebite treatment.
4. Look up snakebite treatments recorded in literature. For example, how did Jody and Penny treat snakebite in *The Yearling*? A librarian or literature teacher should be able to steer you into quite a number and variety of treatments. Compare some of these with the steps recommended by the American Red Cross in their basic first-aid training book.

100. The Law of the Land

What is the difference between a fruit and a vegetable?

If tomatoes were fruit, Mexican farmers could have shipped them into the United States duty-free; if they were vegetables, a tariff was due.

Quite naturally, and quite biologically, the Mexicans maintained their crop was fruit. Customs collectors north of the border viewed tomatoes as vegetables.

The argument wound its way up to the United States Supreme Court.

How would you have found?

The testimony of biologists did not impress the justices.

Because tomatoes were not eaten as a main course component, the nine justices ruled them fruit, and duty-free. It is the law of the land.

Supplemental Activities

1. Trace the derivation of the word *vegetable*, then comment on the linguistic impropriety of referring to coma victims as vegetables.
2. *Vegetable* has no universally accepted, biological definition. Vegetables include roots, stems, leaves, and flowers. Make a list of examples of vegetables from each of these categories. Check your work, because appearances can be deceiving. Potatoes, for example, are *not* roots!
3. Which of these "American" vegetables actually developed in the New World: beets, cabbage, corn, onions, peas, radishes, or tomatoes?

---

101. Ten Months Vacation, Anyone?

What would you do if you only had to go to school for two months a year to learn all you need to learn?

To the American Indians, corn was a gift from the gods. Although scientists trace corn's development back to fertile lands along the Mexican-American border, it is little wonder that the plant seemed a holy gift to the Indians.

About 50 days of work a year with corn fields provided the Indians of Mexico and Central America with more food than they needed and time to develop in other ways.

The Indians did not vacation; life was not that good! But with the time and wealth that corn provided, they accomplished much in mathematics, astronomy, science, agriculture, art, and religion.

Supplemental Activities

1. If by chance you have gotten this far in school without learning about the Aztecs, Toltecs, and Mayans, report to the library immediately for some of the most interesting reading in social studies there is.
2. Find out what each of these plants did for civilization: potatoes, olives, coconuts, and rice.
3. Speculate on the relationship between corn and the native peoples' need for knowledge of agriculture, astronomy, mathematics, and religion.

102. As American as Apple Pie?

How "American" is apple pie?

Stone Age people, the lake dwellers of Central Europe, collected apples to eat fresh or to dry for later use.

Apples, which probably developed in the fertile lands between the Black and Caspian seas, spread with civilization through Europe.

By the fourth century B.C., apples were the most important fruit in Europe.

Considering their popularity, food value, and ease of storage, it is not surprising that apples were brought to the New World by Dutch, French, and English settlers.

Apples, like carrots and a few other imported delicacies, were prized by the Indians, who traded seeds across the continent faster than did the pioneers, let alone Johnny Appleseed.

Although apples are not American in origin, the most popular varieties eaten in the United States are very much American. Besides, "as Central European as apple pie" just doesn't sound right!

Supplemental Activities

1. Speak or write to justify this statement: Apples *are* as American as the American people.
2. Explain this statement: New varieties of apples grow from seeds, old varieties grow from trees!
3. Find the stories behind the development of such apples as McIntosh, Winesap, Granny Smith, Delicious, and Baldwin.

---

103. Backyard Borgia

What is the significance of this rhyme? Leaves of three, leave them be.

A nine-year-old shinning up a tree, a hunter patting his beagle, a lumberman changing a tire on his skidder, a middle-aged woman burning autumn leaves, and a lady weeding her garden all fall victim to the backyard Borgia: poison ivy.

Less than an ounce of oil extracted from 80 pounds of poison ivy bark coupled with 13 years of research revealed the secret of poison ivy to Dr. Charles Dawson: a variable blend of four chemicals.

Reactions to poison ivy exposure vary with individuals and seem to be an allergic response, more intense in some cases than others.

No matter what your past history, beware: repeated exposure seems to result in poisoning, sooner or later. So watch out for the backyard Borgia, a poisonous plant most readily recognized by its three-leafed arrangements.

Supplemental Activities

1. Look up the lac trees of Japan for an interesting side trip into useful plant poisons.
2. Cashew nuts come from a poison ivy relative. See if you can learn which part of the cashew plant is poisonous.
3. A number of vegetables come from plants that are or were believed to be poisonous. Look up this aspect of rhubarb, tomatoes, and potatoes.
4. What part of apples contain cyanide, a deadly poison?

104. Food for the Gods

Which beans did Cortez drink from golden goblets that you can now buy as a flat bar?

The Aztecs called it xocoatl and served it to Cortez in gold goblets.

Cortez found the drink bitter, watery, and considerably less interesting than Montezuma's tableware, but he brought some back to Spain.

In Europe its potential was realized, and the bitter drink of the Aztecs became a rich, sweet treasure.

The plant, *Theobroma,* produces squash-sized yellow fruit, within which whitish beans develop. The beans are fermented, sun-dried, then shipped for final processing into a sweeter beverage than Cortez ever knew and into bars you can buy regardless of your age.

Have you guessed that the beans are cacao and chocolate is the finished product?

Supplemental Activities

1. Find out how chocolate is made.
2. Look into the nutritional value of chocolate—you will be pleasantly surprised even if you are concerned about your weight or complexion.
3. What other botanical treasures were brought back by the conquistadors?
4. The chocolate plant is known as *Theobroma.* Get the name translated from Latin to see if you think the name is appropriate.

---

105. Yucca Yuck!

Why did American Indian children eat their shampoo?

Chewing yucca leaves to free fibers for cord-making was not the favorite job of any of the Indians of the southwestern United States—not until someone discovered, by accident or inspired guess, what would happen if the leaves were baked first.

Yucca leaves baked for a couple of hours in a fire pit become so sweet Indian children begged to be allowed to chew the fibers free.

Yucca as a soap substitute might not have been so thrilling to the youngsters, but bruised yucca roots, briefly soaked, can be used to lather a person's hair just like a shampoo.

Although we get cordage from other plants, yucca is still touted as a shampoo ingredient.

Supplemental Activities

1. Survey conditioners and shampoos to see which plants are used in them, then check to see what these plants may really contribute to our daily hygiene.
2. Books on folklore and Indians will supply information about other plants and their uses in past times.
3. Herbariums are located in museums all across our country. The labels on specimens sometimes include notes about uses of plants by indigenous people. Obtain permission to check such notes. Scientists like Siri von Reis Altschul are reviewing notes such as these to see if important medical applications might be rediscovered.

106. Warm-blooded Plants

The ability to maintain a constant body temperature—being warm-blooded—is a distinct advantage for modern birds and mammals. It is not a characteristic of lower animals, certainly not one expected in plants.

Surprisingly, research has shown that a common North American plant, skunk cabbage, can maintain a body temperature as much as 60 degrees higher than the immediate environment.

How is this possible? How would this serve the plant?

These are the kinds of questions such discoveries usually engender.

By conducting respiration at an incredible rate, skunk cabbage appears to be able to keep a temperature of 70° even when growing through spring snows in the Northeast.

Supplemental Activities

1. Explain how heat would be generated by respiration.
2. Speculate as to how this warmth might assist the plant and what effect it might have on such animals as bees, spiders, and salamanders.
3. Spiders frequently build webs across the oddly shaped flowers of skunk cabbages. What might they be doing?
4. If the skunk cabbage grows in your area, see if you can verify the claim that it generates heat and holds a constant body temperature.

---

107. Bearer of Death, Bearer of Life

What might a monkey dying in the jungle and a heart surgery patient fighting to live have in common?

A monkey shrieks at a sudden pain!

High in the tree, the monkey had suspected nothing until the sliver of wood stabbed into his rump. He starts a panicky flight through the treetops, then falls dying to the ground, the blowpipe dart still imbedded in his hindquarter.

This day a South American Indian will feast on monkey meat.

Half a world away, an anesthesiologist administers a muscle relaxant as part of the regimen of medication necessary for the patient's particular surgery.

This day in a U.S. city, a man will survive surgery and live a normal life again.

The same chemical that killed the monkey saved the man. This bearer of death, bearer of life, is curare.

Supplemental Activities

1. Look up the story of the discovery of the medical applications of curare.
2. Using the *Merck Index* or other reference, find the structural formula of curare.
3. Research the methods South American Indians use to make their arrow poison. In addition to the plants used to make curare, some tribes use the secretions of certain frogs—poison arrow frogs.

108. Lost Health Secrets

"Lost health secret discovered!" reads the tabloid headline. "Wonder diet food actually cancels calories!"

Lost health secrets, wonder diets, cure-alls, and magic cosmetics have been gimmicks for enticing the unsuspecting ever since movable type was invented.

What are the chances of lost health secrets being rediscovered?

Such claims as those quoted above are recognizably bogus by their very wording, yet the possibility exists that lost health secrets will actually be rediscovered.

Peasants, aboriginal peoples, and just plain country folks used local plants in varied ways. As modern science offered easier substitutes, the old ways were forgotten. Valuable secrets, clues to plants with real medical value, may lie overlooked in the botanical collections that have been piling up in museums since the 16th century.

Notes on such specimens recording uses by indigenous people may yet lead to important medical discoveries.

Supplemental Activities

1. Look up these medicines to see where they come from: ipecac, curare, digitalis, and belladonna.
2. Interview elderly people about their uses for plants and other folk remedies.
3. Although cocaine is now recognized as a dangerous "recreational" drug, its medical uses and the uses of various synthetic versions of cocaine are well established. Find information about the use of coca leaves by Indians of South America.

CAUTION: DO NOT try to treat yourself with wild plants or folk remedies.

---

109. Love Apples

What "poisonous" plant is the main ingredient of America's favorite condiment?

In Colonial New England, love apples were planted as an ornamental. Their lovely red fruit graced many a garden, but no one's table. It was widely known that love apples were as deadly as they were beautiful.

Imagine a young man so distraught to find his lover unfaithful that suicide seems his only path to peace. He sits down in the garden, plucks the reddest love apple, eats it all, then waits patiently for death despite the pleasant flavor of the *tomato*!

Supplemental Activities

1. As amusing as the notion of attempted suicide by tomato may seem, there is a seed of truth to the idea. Look up the family to which tomatoes belong to discover why the idea of poisonous love apples is not so outlandish.
2. Although tomatoes first grew in the New World, they came to North America only after traveling about much of the world. Research the story of the tomato from Mexico to catsup bottle.
3. Is the tomato a fruit or a vegetable? For those who like to dig, find the Supreme Court decision that settled this question.

110. Thistle Thou Never Wort

Think of Western movies you have seen. Imagine a ghost town. The wind whistles through chinks in wooden buildings. A shutter bangs. Dust devils dance in the street.

What plant blows down the street, as much a part of Westerns as John Wayne?

Sometime after the American Civil War, the first Tartar thistles took root in American soil. Their seeds were probably stowaways in a shipment of flax from Europe.

The Tartar thistle, actually a saltwort, was well adapted to the barren alkali lands of much of the American West.

Americans were quick to rename the Tartar thistle, commemorating its peculiar method of spreading its seeds: the whole above-ground part of the plant breaks free and is blown about as the seeds are shed. Now thought of as an American plant, the thistle is famed in the West as the tumbleweed!

Supplemental Activities

1. Collect and examine a tumbleweed.
2. Research the reproductive strategy of the tumbleweed.
3. Try to find the European range of this plant.
4. Several not-too-successful attempts have been made to turn this nuisance weed into a commercial product. Find out about these attempts. Perhaps you will be the one to find a way to turn tumbleweed into money and some good use.

---

111. Gift of the Jaguar God

Which "all-American" vegetable once inspired human sacrifice?

High in the Andes, the Indians thanked the Jaguar God, the Creator, for potatoes by making blood sacrifices.

Once discovered by Europeans, potatoes were regarded as a food, of course, but also as an aphrodisiac, a cure for rheumatism, a cause of leprosy, and a source of poison.

Of those four odd beliefs, one is actually true. Raw, green potatoes contain an alkaloid, poisonous until cooked.

Supplemental Activities

1. Trace the routes of potatoes from South America to North America.
2. British soldiers, bent upon punishing Irish revolutionaries, waited until harvest time, then burned the potato fields. Why did this practice amuse the Irish farmers?

112. The No-Sting Plant

When the little girl threw a rock through the hornet's nest, the old fishing guide assured the anxious parent and the squalling child that this didn't have to end the morning's fishing before it even got started.

"When I was a young 'un," the guide explained, "an Indian medicine man showed me the 'no-sting plant,' and I'm going to show it to you."

"A real medicine man?" the girl stopped crying long enough to ask.

The old man nodded.

The guide stepped off the brook trail and pulled a number of succulent leaves off a tall plant with lovely orange flowers.

"Will it work?" the girl's father asked.

That's the question. Will some brookside plant ease the stings of bees, or is the old man just applying a bit of psychology?

Nodding, the guide crushed the leaves of the jewelweed in his rough hands.

Soon, the girl was fishing happily, the leaves wrapped to her arm with her dad's bandana. Her painful sting was forgotten—not even a bump left. There was no doubt in the girl's mind that jewelweed is indeed the no-sting plant.

Supplemental Activities

1. Jewelweed is one of a number of plants held in high regard by the Indians as medicinal. Can you find out what active ingredient(s) are present in jewelweed sap? What other medicinal uses does jewelweed have?
2. Question a doctor about the psychological effects of medicine. You may be surprised to hear the answer.
3. Jewelweed is interesting because of the unusual way it distributes its seeds. Most botany books use it as an example of seed dispersal. Check this one out—in the field if possible. How does it relate to jewelweed's other common name?

Fossil Fuel

113. Dr. Gridley's Tar Medicine

Dr. Gridley peddled his Tar Medicine from a brightly painted wagon over a wide circuit of small towns in Maine a century ago.

It was a panacea, a cure-all—according to the "doctor," whose claims were as bogus as his degree, both products of a man's enterprising imagination.

What kept such phony medicine peddlers in business? How can quacks succeed with a worthless product?

It is probable that the alcohol in the "medicine" contributed most of the effects and perhaps led to repeat sales; it is certain that tar accounted for the medicinal taste.

Whether or not the Tar Medicine did anyone any good cannot be known. However, in many a temperate household, whatever solace might be found in drink was found in Doctor Gridley's Tar Medicine—bought and consumed without fear of censure.

Supplemental Activities

1. "Tar," crude petroleum, was used in medicine long before one of the author's ancestors decided to take his product to the people of New England. Look up medicinal uses of "tar" past and present.
2. In the Holy Land, "pitch" is also a petroleum product. Check the Bible for references to the uses of pitch in ancient times. The book of Genesis is a good starting point.
3. Long before an eternal flame was installed to mark the grave of the assassinated President John F. Kennedy, travelers in the Arab world encountered great pillars of fire burning in the deserts. Explain this phenomenon, and then confirm your hypothesis by library research.

Fungi

114. Ritualistic Urine Drinking

Many years before the Russian Revolution, a group of Chukchee tribesmen gather around a campfire, a speck of warmth in the icy vastness of Siberia.

A lone man dances wildly about the fire. The others watch and wait as the dancer leaps, gestures, and twists. The dancer sings of things he and he alone sees: commonplace objects many times life-size; bright, constantly shifting colors, brighter than the aurora; and fanciful shapes.

The dancer pauses to urinate into a wooden bowl.

From person to person, the bowl is passed. Each person drinks from the bowl!

What could lure one man to drink another man's urine?

After a bit, all are dancing about the fire, making the night wild with their antics.

They are sharing, in a most unusual way, the hallucinogenic chemicals of the agaric mushroom the first dancer had consumed. The chemicals passed through his system unchanged, making his urine as psychoactive as the mushroom itself.

Supplemental Activities

1. Research the hallucinogenic and poisonous properties of fly agaric, *Amanita muscaria.*
2. Report on the toxic nature of human urine.
3. Find out about other ways people have consumed hallucinogenic drugs. Teonanacatl, virola powder, and kwashi would be good drugs to look up, for their methods of consumption are unusual to say the very least.

115. Witches'-Brooms

What disease threatens your chocolate bar?

You will see them occasionally, odd clumps of small twigs, densely packed, arising from the same branch. They are a deformity caused by a parasitic fungus: witches'-brooms.

Witches'-brooms, as their name suggests, have a history in folklore.

Their presence might force up the price of your chocolate candy.

A witches'-broom that attacks Brazilian cacao trees has reached epidemic proportions in much of the Amazon basin's chocolate-producing areas.

Attempts to confine it to one region have been moderately successful. However, if it should spread—especially if it should spread to such other major producers as the Ivory Coast—the results could be disastrous to chocolate merchants and chocolate lovers.

Supplemental Activities

1. Collect samples of witches'-broom from local plants.
2. An agricultural extension service may provide information on witches'-brooms and other fungi that threaten local plants.
3. Make up a folk tale to explain witches'-brooms.

---

116. A Reproductive Strategy

"What is that stench?" the camp counselor asked accusingly.

The twelve-year-olds in his tent, although not known for making places smell good, were innocent of this creation.

Together, the man and the boys went through the tent, beds, footlockers, and laundry bags. Still the smell was a mystery.

Outside the tent, they found the culprit: an eight-inch-high fungus. The stinker looked like a mushroom stem devoid of any cap.

"It smells bad enough to gag a maggot!" one boy exclaimed in disgust.

"That smell," the man explained, "helps this organism reproduce!"

How could a foul odor help a fungus reproduce?

Actually, the camper's reference to maggots was right to the point. The fungus, known as a stinkhorn, smells bad enough to attract flies in search of food or places to lay eggs. The insects fly away with fungus spores on their legs, helping the fungus to reproduce.

Supplemental Activities

1. *Pilobolus* is another fungus with a downright spectacular reproductive strategy. Check it out. You might even grow this fungus in the lab.
2. Find out which carnivorous plant uses garbagelike smells to attract flies.
3. Find a ripe puffball, one that "smokes" when touched. Study some of the smoke particles, actually spores, under a microscope. Can you induce them to germinate?

117. Attack of the Micromonsters

Below the surface, where the soil was damp, the tiny roundworms swam easily in the film of moisture on rock and humus particles.

Some, like unwary soldiers on patrol, would fall victim to booby traps—living booby traps called predaceous fungi.

One hapless roundworm slid through the open part of a doughnut-shaped fungus.

The roundworm was doomed. What would happen before the worm could pass through the "doughnut"?

At the slightest touch of its inner surface, the fungus suddenly ballooned, strangling the worm in the shrinking opening. Hyphae, specialized feeding cells, invaded the tiny, squirming victim, digesting it from within.

The micromonster *Arthrobotrys* had claimed another victim.

Supplemental Activities

1. Look up another predaceous fungus such as *Dactylella* to learn of other living booby traps.
2. How do fungi usually feed?
3. Using balloons, demonstrate the trapping system of *Arthrobotrys*.
4. How might osmosis account for *Arthrobotrys*'s ability to swell quickly enough to trap soil nematodes?
5. Find out how changes in water pressure account for motion in starfish, Venus's-flytraps, and bladderworts.

118. An Interrupted Lunch Becomes a Feast

What improves with mold?

Every day the boy drove the sheep to pasture, guarded them, then drove them back to the village for the night.

It was a responsible job for a boy and one he did with pride.

Like most peasants, this French youngster ate simple fare indeed: coarse bread, soft cheese, whatever greens might be in season, and meat when available.

So it was on one hot day, while the sheep grazed peacefully. The boy sought the shade of a cool cave to eat his lunch.

Terrified bleats and snarls no sheep could voice interrupted the boy's lunch. Dashing out of the cave, he came face-to-face with a huge wolf. Without thinking, the boy picked up a rock, threw it at the wolf, and killed the predator instantly.

The boy hurriedly urged his sheep back to the village, where he displayed the wolf's ears and basked in the praise heaped upon him.

Weeks later, when by chance he brought his sheep back to that particular pasture, the boy ventured into the cave. His cheese remained where he had left it, but not unchanged. Mold had invaded it.

Mustering his courage, he tasted the moldy cheese. One more time he hurried back to the village. The cheese was so improved by the mold that the people of that region have continued to cure their cheeses in the caves of Roquefort.

Supplemental Activities

1. This story is best regarded as legend, but so are many of the stories of ancient discovery. Locate some of these other legendary discoveries and share them with the class.

2. What does cheese contribute to our diet?

3. Comment upon this: The diet described in this opener is not varied, but neither is it unbalanced.

4. Could you make cheese in the school laboratory?

CAUTION: DO NOT sample moldy foods in hopes of making the next great culinary discovery. Many molds are **unsafe** to eat.

119. Rangers' Night Light

If you had no flashlight, could strike no match, nor build any kind of fire, how would you light your way on moonless nights in gloomy forests?

During the French and Indian War, Major Robert Rogers led a raid against the St. Francis Abenaki Indians. Rogers force-marched his men day and night through hostile territory.

In the gloomy forests, when a lantern (let alone a fire) would have betrayed their presence to the vastly superior forces of their enemies, the rangers had a natural light. A light that would pass unnoticed if seen by the enemy.

Rogers read his maps by the ghostly, blue-white glow of a woodland mushroom's mycelium.

Had either French or Indians seen this pale light, unlikely at best, the luminous fungi would not have seemed out of place.

Supplemental Activities

1. Finding and experimenting with luminous fungi seems the most natural supplementary activity. A little time with books on regional fungi and some poking about in spooky forests should prove rewarding.
2. Look up bioluminescence to see how this cold light works and to learn what a variety of organisms have their own biological night-lights.
3. In season, capture and observe fireflies, the most common bioluminescent organism over much of North America.

---

120. Fungi at War

What is jungle rot? creeping crud?

The soldiers called them jungle rot and creeping crud. They were the diseases caused by fungi that plagued soldiers in the Pacific during World War II.

More than half of the American soldiers hospitalized in the Pacific Theater were struck down by fungi, not Japanese forces.

Except that the fungi have no racial prejudices, no politics beyond survival, and parasitize Orientals as willingly as Occidentals, these organisms might have tipped the scale of war in favor of Japan.

Less than half the goods shipped to soldiers on Pacific islands arrived undamaged by the ubiquitous fungi.

Supplemental Activities

1. You might be interested in finding out how fungi affected the course of other wars.
2. What names did the American forces in Vietnam invent for their fungi-based diseases?
3. What can fungi eat? Or, perhaps better asked, what can't fungi eat? This list will help you understand how adaptable and therefore how potentially bothersome the fungi are.

Genetics

121. Sheep to Count

Early in the summer of 1791, Seth Wright, a farmer in Dover, Massachusetts, strolled into genetic fame.

As Seth Wright walked his fields, he enjoyed watching new lambs frolicking in the soft grass or nursing contentedly. It was more than seeing his flock increase that brought a smile to his face. But one lamb brought forth outright laughter. It was the funniest-looking little ram Wright had ever seen. The little fellow's legs were so absurdly short his belly wool dragged as he trundled along.

Before summer, however, Seth Wright saw a real advantage to short-legged sheep — not for them, but for him.

How could short-legged sheep be better for the shepherd than normal ones?

The stubby-legged ram could not jump the stone walls that separated Wright's fields from those of his neighbors. Seth reasoned that a whole flock of short-legged sheep would be far easier to herd. So he decided to breed the little ram.

Given 15 ewes, the little ram participated in farmer Wright's experiment wholeheartedly! From that ram's progeny came a whole, true breeding race of short-legged sheep.

This was the first germinal mutation of domestic animals in the New World.

Supplemental Activities

1. Look up the stories of navel oranges, seedless grapes, and Delicious apples.
2. Speculate as to how short-legged sheep would have fared in the wild.
3. How can seedless fruits and vegetables be perpetuated, since they cannot produce new generations by seeds or pass on their genes in the usual manner?
4. Comment upon this statement: Mutation is the good news; mutation is the bad news.

122. DNA from Ancient Brains

How might we use the DNA of our ancient ancestors if we could obtain samples of it?

The girl's death brought great sadness to the tribe. Children held a special importance for these people. Her grave would be filled with gifts: trinkets she had liked in life, tools she would need in her next life.

Her body, laid out as if she were sleeping on her side, was held in place under the water of the bog by stakes. Durable, well-woven cloth and a grass mat covered the child.

There she lay, in a Florida bog, for nearly 8,000 years!

Although her skin, muscles, and internal organs decayed, her small skeleton endured as, most incredibly, did her brain!

The child's brain, shrunken and discolored, was so well-preserved that tissue samples including the DNA, the genetic code that had made her a beloved child, are intact: a treasure for scientists today as she was a treasure to her people so long ago.

Supplemental Activities

1. In the movies, the scientists might clone her cells, revive them, and create a teenybopper-sized Frankenstein monster. What do you suppose real scientists will try to learn from these most ancient samples of human DNA?
2. In the "dark casket" of peat, preserved remains of humans have been found in Denmark, Ireland, and other ancient European bogs. Research these "bog people" to see what we have already learned from them.
3. "The peat is the dark casket, where we have found many clues to our past and to our cultural identity." This is a quotation from Seamus Heaney, an Irish poet, not an archaeologist or geneticist. Discuss how poets, visual artists, and writers might view such finds. Then seek out some of Heaney's writings to compare with your predictions and expectations.
4. Try your own hand at composing some work of art—music, dance, poetry, sculpture, painting, or photography—that will reflect what you have learned from study of the bog people and what feelings knowledge of them evokes in you.

123. Flowers of Death

What would happen if some very important plant, a plant whose products were vital to humankind worldwide, died—worldwide, in inexplicable mystery?

The death of the bamboo was a tragedy for the villagers along the little jungle stream in Borneo.

Bamboo supplied them with building material, tools, fiber, food, shelter. It was a vital part of their world, a part of their lives. Now all the bamboo was dying.

Grainlike seeds weighed down the slenderest branches. The tall, graceful culms (stalks) were dying—in their villages, throughout neighboring villages, throughout the world!

At nearly the same time, each species of bamboo, wherever it is found in the world, flowers and seeds and then dies back to underground stems called rhizomes.

As often as every 30 years or as rarely as once every century and a quarter, depending upon species, each kind of bamboo dies when its allotted time is spent.

Programmed into the genetic code of these largest of all grasses is this peculiar pattern: live, reproduce, die—then live again.

Supplemental Activities

1. How might mass death of adult plants be an advantage for the species?
2. Research other organisms that die after reproduction. You can find examples throughout the five kingdoms of life.
3. Report on the uses of bamboo. The *National Geographic* of October 1980 or an encyclopedia would be a good starting point.
4. Find articles made from bamboo for classroom display. Fly rods, decorative baskets, canned "Chinese" food, even living specimens are among easily obtained examples. Many others are available to the creative, resourceful searcher.
5. Why hasn't bamboo become extinct?

Gymnosperms

124. Sole Survivor

What tree loves and thrives in the polluted air of industrial cities?

During the Permian Period the beautiful ginkgo trees first began to grace the landscape.

Survivors of the Permio-Triassic Crisis which wiped out most of the life forms on Earth, the ginkgo trees prospered throughout much of the Mesozoic Era. Dinosaurs played out their whole existence in a world of ginkgo forests.

Whatever fate befell the dinosaurs ended the Mesozoic Era and greatly reduced the surviving species of ginkgo. By the 10th century A.D. only one species remained. Had Chinese monks not tended the trees in temple gardens, they might not have become so familiar a part of our cities.

Able to endure smog, fumes, pollution, and the less-than-ideal soil conditions of cities, the ginkgo tree is the hardy sole survivor of ancient forests.

Supplemental Activities

1. Might the ginkgo have success in cities because the atmosphere there is like that of ancient times? Check out that possibility.
2. Collect samples of ginkgo trees: leaves, fruit, bark, etc.
3. Trace the life cycle of the ginkgo — an interesting theme and variation on alternation of generations and a glimpse at another step in plant evolution and mastery of life on land.
4. Find out why male ginkgo trees are more popular than female.

125. Leafless Wonder Plant

As the Mormons crossed the desert on their way to Utah, they could not help but notice a peculiar plant: a scruffy, leafless, green-stemmed shrub.

The Indians told the Mormons the shrub was medicinal, bringing relief when brewed as a tea to those who had difficulty breathing.

Medicinal tea was acceptable to the Mormons. They greatly respected the wisdom of the Indians, so they used the tea to treat those of their number who suffered from respiratory problems. The tea worked.

What medicine do we still extract from Mormon tea?

The plant, known to science as *Ephedra,* is now a source of a powerful medicine, ephedrine. Many who suffer from hay fever, asthma, and other respiratory conditions get real relief from modern medicines made from Mormon tea.

Supplemental Activities

1. By looking up digitalis, folk remedies, or medicinal plants, you could track down stories of many other interesting medicines extracted from plants known to Indians.
2. Check medications and patent medicines to see how many contain ephedrine.
3. Ask a representative of the Church of Jesus Christ of Latter-Day Saints (the Mormons) to tell you other things the Mormons learned from the American Indians.

---

126. The White Bread Plant

The Seminole Indians of Florida's Everglades make sago. So do unrelated peoples on four other continents: people of five worlds making the same product from different plants.

What is sago?

For one thing, sago is a discovery made independently in different parts of the world by humans who were unaware of each other.

Sago is also a way to use an otherwise inedible part of several similar but unrelated plants, including cycads and palms.

Each of the sources of sago is a palm or palmlike cycad: plants with coarse, fibrous pith in their stems. People all across the world learned to dig out the pith, grind it, wash away the poisons and remove the fibers in water, then dry the remainder into granular starch eventually to be made into pudding, unleavened bread, even starch for cloth.

Sago from a cycad was the source of the white bread of the Seminoles, who were known to have no wheat or similar grain for flour yet broke bread for their daily meals.

Supplemental Activities

1. Look up sago and its manufacture for an interesting and important food story.
2. What other foods have come to us from American Indians?
3. Trace the development of flowering plants from their gymnosperm ancestors, with special attention to the theoretical role of cycads.
4. Find pictures of Mesozoic life, especially dinosaurs, and identify the cycads.

127. Impossible N'tumba

What does a plant look like?

Think about it for a moment, perhaps from the perspective of describing a plant to an intelligent being who had no idea what plants were like.

Lured by the legend of the n'tumba, guided by bushmen, Fredrich Welwitch found himself deep in Africa's Mossâmedes Desert.

Huddled around a fire of animal droppings, more as a protection from night chills than any wild beasts, the bushmen told their Austrian boss more about n'tumba—a plant unlike anything Welwitch had ever known.

What plant could be a scant foot high, yet have two 10-foot-long leaves and a root too deep for men to dig?

The more detailed the bushmen's description, the more inclined Welwitch was to think himself the victim of a prank.

Before the safari left the desert, Welwitch learned that his guides spoke truly.

With two great, straplike leaves growing out of a trunk little more than a stump, and a single taproot as much as 80 feet long, the existence of the plant, n'tumba, now called *Welwitchia,* does seem impossible.

Supplemental Activities

1. Explain how the design of n'tumba is adaptation to desert environment.
2. How did the bushmen use n'tumba? Several interesting stories lie within the answers to that question.
3. What other plants come from the Mossâmedes?

128. Jungle Man, Jungle Vine

Raise your hand when you think you know the name of the famous author I am describing.

Born in Chicago in the late 1800's, he would try a variety of jobs with little success until he turned to writing.

In 1912 he wrote a story about a boy raised by African mammals.

He would write 67 books—36 about his jungle person, many others about adventures in space.

The jungle-man books would be translated into 26 languages and serve as the basis of a long-running comic strip, more than three dozen movies, two television series, and a long succession of comic books.

His Lord of the Jungle, Tarzan of the Apes, would, in the movies, swing from tree to tree on jungle vines.

Who was the author?

Edgar Rice Burroughs.

The vines at least are real. Burroughs knew them from descriptions and pictures. They are lianes.

Lianes are odd gymnosperms, odd because unlike most living gymnosperms, they are broad-leafed. Too rigid to swing on, they are the vines that festoon trees Burroughs saw in pictures of African jungles as he did library research for his books.

Supplemental Activities

1. Read books or view movies such as *Stalking the Wild Child* to explore legends of feral children.
2. Find some other broad-leafed or leafless gymnosperms in biology or botany texts.
3. Are there some odd gymnosperms indigenous to your area?
4. Compare the jungle of Tarzan movies with the real jungles of Africa.

Insects

129. Of Beetles, Gods, and Camel Droppings

What are scarabs?

To the ancient Egyptians, the scarab, a dung beetle, was sacred.

How did it come to pass that the Egyptians would associate a beetle that eats camel droppings with their most powerful god, the sun god?

The beetle forms a ball of manageable size from the large droppings of camels. Then it trundles off, in reverse, pushing the dung ball to a convenient place for burial. The beetle, a female, lays her eggs in the dung ball before burying it.

The maggots hatch underground and feed upon the dung as they develop.

To the ancient Egyptians, the beetle pushing the ball of dung along was the embodiment of the great, unseen sun god who propelled the sun across the sky.

Although one might doubt the Egyptians' grasp of the material of which stars are made, there can be no question as to the value of insects that recycle organic material such as dung.

Supplemental Activities

1. What would happen if animal droppings were not recycled? In *Scientific American,* April 1974, D. F. Waterhouse traces the damage to Australian cattle lands caused by unrecycled dung. (See Opener #75.)
2. Lucretius, who died in 55 B.C., said it first: "One man's meat is another man's poison." Discuss this in relation to the dung beetle, then to as many other forms of life as your interest carries you.

130. Arctic Killers

"TUNDRA VISITOR BITTEN TO DEATH!"

If you read a headline like that, what killer would you suspect?

Would you suspect polar bears? wolves? berserk Eskimos? killer whales?

None of those are so likely as mosquitoes!

In the arctic summer, swarms of mosquitoes surround travelers like living fog. Unprotected people can suffer more than 7,000 bites a minute! That is enough to cause death in a few hours.

The scarcity of humans in that region keeps such headlines hypothetical, but the potential is there.

The dearth of warm-blooded victims presents very real problems to female mosquitoes, who must have protein-rich blood to reproduce efficiently.

Supplemental Activities

1. Only female mosquitoes suck blood, then only after mating. What do males and nonreproducing females eat?
2. How did Eskimos and northern Indians protect themselves from mosquitoes?
3. Blood lost to mosquito bites is far less a risk to humans than diseases acquired from mosquitoes. For some interesting reading, look up these mosquito-borne diseases: malaria, dengue fever, yellow fever, and encephalitis.

---

131. Mesophyll Miners

Some people eat only the filling of their sandwich. Many a youngster eats only the filling of Oreos.

What insect eats only the middle, food-making layers of leaves, leaving the upper and lower epidermal layers undisturbed?

Leaf miners are insects grouped together by habit, not by relationship. Larval stages of several unrelated moths, beetles, flies, and wasps burrow in the mesophyll of leaves, protected from a hazard-filled world by the leaf's protective outer cells.

Discolored trails and patches betray the miners' presence.

Supplemental Activities

1. Collect samples of leaves showing damage from leaf miners.
2. For a real project, find a tree or bush infested with leaf miners and design a study to explore the relationship among the tree, the miners, and perhaps other organisms.
3. Read "Eating a Thin Line," an article in the February 1988 issue of *Natural History*, for an excellent description of ongoing research into leaf miner behavior and effects.
4. Survey a field guide to insects to learn which indigenous insects have miner stages as minors.

132. Pest Paradises

Imagine all of your favorite food laid out for you to eat whenever and as much of as you choose. For some people, that would be paradise.

How does modern agricultural practice create a paradise for pests?

To make agribusiness possible, we have taken plants that might grow in small areas or widely scattered among other species and have concentrated them into vast fields of only their kind. Thus we have set out a feast for insects specializing in those plants.

Supplemental Activities

1. Why have we set ourselves up for crop destruction by insects?
2. For many years, insecticides were the answer of choice to the problem of crop destruction by insects. In consideration of the fact that less than one percent of insects are crop destroyers, why is the wholesale poisoning of insects unwise? Hint: Ask yourself what the other 99 percent of insects are up to.
3. Insects destroy 10 percent of our crops worldwide, but you are being asked to support this statement: Without insects, we would have much less to eat.

---

133. Bold Deceivers

Ants are formidable enemies. To combat they bring poison bites and stings, powerful jaws, well-protected heads, great numbers, social organization, and legendary strength.

Besting ants would seem impossible for other insects.

How could insects of comparable size prey upon ants with acceptable risks?

European *Nitidulid* beetles detect the chemical trails that guide ants and mimic their behavior, thus tricking ants into feeding them as they would their own kind.

Staphylinid beetles ambush and kill ants. If the staphylinids themselves are attacked, they offer the ants sweet secretions from special abdominal glands and escape while the ants enjoy their ransom.

Lomeschusa beetles smell so good that ants allow them to live in their nests even though the larval beetles devour ant larvae.

The beetles that best ants are bold deceivers indeed.

Supplemental Activities

1. The beetles described above are European. Consult insect and ant books to see if such bold deceivers live on North American soil.
2. Discuss the bold deceivers of history. To start you off, consider the Greeks who built that fabled wooden horse.
3. Learn about protective coloration and mimicry used by many kinds of insects as well as other animals for protection.

134. The Flying Frankenstein

"My caterpillars have escaped!" Leopold Trouvelot told the police and the press. "My caterpillars have escaped!"

Neither the police nor the press took much notice of the man with the funny accent and a fear of caterpillars.

Do you know or can you imagine why the escape of Leopold Trouvelot's little French caterpillars was a problem worthy of police and press attention?

In 1869 Trouvelot attempted to bring a new industry to Medford, Massachusetts: silk production. Knowing that silkworms, the larval stage of an oriental moth, did not do well in cold climates, Trouvelot was trying to crossbreed them with the hardier but closely related gypsy moth.

The escape of the gypsy moth caterpillars, sailing away on silken threads, was a disaster.

Far from their native France, without natural enemies, these highly destructive insects would soon establish themselves and begin stripping trees in the forests of the Northeast. Trouvelot's worst fears were realized. His experiments had produced a flying Frankenstein monster which still ravages the forests.

Supplemental Activities

1. Read Mary Shelley's classic novel *Frankenstein* to see how closely these stories parallel each other.
2. Alien insects and other forms of life often do very well for themselves at our expense. To see proof, read about other organisms brought from one environment to another only to escape and run amok: killer bees, English sparrows, and rabbits in Australia would be good starters.
3. What is gypsyl and how are we using it against the gypsy moth?

135. Amazon Slave Raiders

Is slavery a uniquely human institution? Are any other species clever enough, ruthless enough, or mean enough to condone slavery?

"Amazon slave raiders" sounds like the title of a late, late movie, a jungle jumble, from which neither actors nor viewers would make much profit.

Amazon slave raiders is actually a description of ants common to much of North America and Europe.

Called Amazons because they belong to a female society of warriors, *Polyergus* ants raid the nests of other species, kill all adults who resist, and take captive the pupae of their victims.

When pupae hatch, the captive ants become slaves for the *Polyergus*.

The red *Polyergus* ants have become so specialized, so dependent, they cannot survive without slaves. The mandibles of the *Polyergus* warriors are saberlike, fine for combat but useless for feeding. It is the captives who will feed and tend their female captors.

Supplemental Activities

1. Find out how ant colonies wind up being almost 100 percent female.
2. After studying ant slavery further, compare and contrast it with slavery as humans practiced it.

---

136. No Choice of Jobs

How many different jobs are available for your choice?

The United States Department of Labor identifies 40,023 different jobs you could choose as a career. Each of these jobs is open to anyone who is qualified regardless of sex, race, religion, ethnic origin, or place of residence.

For societies far older than ours, no such choices exist. In a termite hill that may have housed more generations of insects than Earth has had of humans, individual occupations are determined by heredity and environment.

There are four basic jobs, one of which is usually monopolized by a single individual. The largest number of termites are workers or soldiers. The queen, the only fertile female in the entire hill, is engaged in the continuous production of workers and soldiers. A few males are produced only to service the queen.

In insect societies, there is no choice of jobs.

Supplemental Activities

1. After studying both, compare the caste system of India with the caste system of social insects.
2. What distinguishes social insects such as ants, termites, and bees from solitary insects such as houseflies, mosquitoes, and water striders, which may live in large populations?
3. Trace the evolution of social insects.
4. Explain why social insects are so successful.

137. Computer Bugs

What are computer bugs?

Excepting cartoons and the movie *Tron*, computer bugs are system problems, not organisms.

Several mainframe computers, however, have been damaged or crashed by real bugs—insects.

To anyone familiar with insects, it will come as no surprise to learn that the offending insect is humankind's old bugaboo, the cockroach.

Supplemental Activities

1. Look into the troubled relations between humans and cockroaches.

2. For an interesting sidenote, research the geohistory of the lowly cockroach.

3. How could a mainframe computer be debugged without risking damage to the hardware?

4. In all probability, cockroaches are available to you. After checking safety precautions and being sure it violates no laws or ordinances, raise cockroaches in the laboratory and study their behavior.

---

138. Life Close to Death

What opportunities for success could insects find in insect-eating plants?

The ratty little plume moth is an insect easily overlooked. Plume moths are small, even as moths go. Their wing fibers fall off easily. Yet they are beautiful in their adaptation and ability to live close to death.

As larvae, the bristly little plume moth caterpillars forage for food on the leaves of sundews, small but efficient insect-catching plants.

Like soldiers feeling their way through a mine field, the plume moth larvae move between the sticky, insect-catching spikes of the sundew leaves. The caterpillars' long spines keep them from being caught.

The caterpillars eat only the central spines of the sundew, leaving plenty for the plant and enough to protect themselves from insect predators.

The larvae pupate and grow to maturity on sundew leaves.

Adult plume moths may return the favors of safe childhood and food by pollinating the sundews.

Supplemental Activities

1. Further reading on the plume moth will tell you how the adults avoid death by digestion.
2. Look up other "carnivorous" plants to learn how they have become home to animals who can adapt to life close to death.
3. Collect and raise sundews or other carnivorous plants.

139. Yellow Brick Road

How do tent caterpillars find their way from home to food and back again?

Enlivened by warm morning sunlight, the tent caterpillars stir within their silken shelter. First an adventurous few emerge, then more and more trundle off to satisfy their only interest: hunger.

Several times during the day, the caterpillars return to the nest only to go foraging again. As the day progresses, larger and larger numbers of the immature insects concentrate where food is most abundant.

The caterpillars lay down a trail of silken threads and stimulating chemicals. The most heavily marked trails are recognized as paths to the best feeding.

Supplemental Activities

1. If tent caterpillars are common in your area, set up an observational experiment to test the conclusions implied here.
2. Scent trails are used by quite a few insects and other animals. You might set up an experiment with some other animals, ants for example, to see if they use scent trails.
3. Why is returning to the nest regularly an important part of this food-finding behavior?

---

140. The High Marsh Gambit

There is no question about it: Fish eat baby mosquitoes. However, mosquito eggs must hatch and baby mosquitoes must live in water.

How can mosquitoes rear their young free from menacing fish?

Mosquito eggs laid in fish-infested waters are at risk. This is especially true in estuaries and salt marshes, the nurseries of many kinds of fish.

The saltwater mosquito has a stratagem: the high marsh gambit. The females lay their eggs in *dry* depressions above the normal tides. Rain or spring tides supply the water required for hatching and larval development, and this water is free of pesky fish.

Unfortunately for all concerned, fertile female mosquitoes require blood from warm-blooded organisms for maximum fertility. Their biting has, to understate matters, brought them to the attention of humans and made them a nuisance for people living along the shore.

Supplemental Activities

1. As a state conservationist, you have been assigned to control the marsh mosquito without using chemicals that might harm the many valuable organisms that abound in salt marshes. Plan a strategy that will limit mosquitoes without endangering the marsh ecosystem.
2. Research the life cycle of mosquitoes to see exactly why the females bite and how the males survive without blood.
3. How are mosquitoes and other bothersome insects controlled in your area? Are these methods environmentally sound?

141. Water Pennies

The force of moving water is surprising. The Appalachian Mountain Club reports incidents in which healthy, strong hikers drowned in only a few inches of water, because they were unable to free trapped feet and stand up as the water pressed them down.

How can the apparently frail insects that inhabit fast-flowing brooks cope with such force?

In upstate New York, a brook cascades down ledges and over rounded boulders with such force it would seem to keep the rocks forever swept bare of all living things. Such is not the case.

The water pennies cling to the rocks and slowly move about in search of food. Resembling ancient trilobites, their upper shells are low domes over which the water rides. Underneath, six little legs prove the creatures' relationship to insects, not crustaceans. Bristles fringing their shell edges secure the water pennies to the rocks. Like the judo master who uses his opponent's strength to defeat him, the water pennies use the force of the water to their advantage.

Water pennies are immature riffle beetles who, as adults, will use other skills and adaptations to wrest their livelihood from fast water.

Supplemental Activities

1. Good designs for living appear over and over again in different, sometimes unrelated, creatures. That water pennies superficially resemble trilobites is a testimony to a good design but not to either coincidence or relationship. Find other examples of animal look-alikes.
2. Parallel evolution is the term for the development of different organisms with similar characteristics: birds and bats, for instance. As a follow-up to your work on Supplemental Activity #1, pursue an understanding of parallel evolution further through reading and discussion.
3. How are trilobites and water pennies alike? How are they different?

142. Indefatigable Jumper

If you could jump as well as fleas do, to what height could you jump?

The flea had hidden in the carpet for most of the week since it last fed.

Vibrations from the oncoming dog alerted the flea. It crouched to leap, more from a kneeling than a standing position, then sprang nearly straight up. At the top of its jump, it was more than a hundred times its body length above the carpet—the equivalent of a person jumping over a building eight stories high!

All six legs (complete with bristly hooks) widespread, the flea was a biological grappling hook. It caught dog hair and burrowed down to bite.

Had the flea missed, it could have jumped repeatedly in pursuit without fatigue. Resilin, a springy protein, rather than muscle powers fleas' legs, supplying both incredible power and stamina. Fleas are virtually indefatigable jumpers.

Supplemental Activities

1. One flea expert reported acceleration in excess of 140g ($g = 9.8\ m/s^2$) in jumping fleas. Compare this with the acceleration of race cars, bullets, and rockets.
2. Research the working of resilin.
3. After reading "Fleas: The Lethal Leapers" in the May 1988 *National Geographic,* discuss the accuracy of these quotations concerning fleas:

   a. "The flea, though he kills none, he does all the harm he can."

   John Donne

   b. "He that lies with the dogs riseth with fleas."

   George Herbert

   c. "If you go long enough without a bath, even the fleas will leave you alone."

   Ernie Pyle

   d. "They say a reasonable number of fleas is good for a dog—keeps him from broodin' over bein' a dog."

   Edward Noyes Wescott

4. Look into the athletic accomplishments of other insects such as ants, water boatmen, and grasshoppers.

Jawless Fishes

143. Wrinkle-Faced Slimer

The first vertebrates lacked a very important skeletal unit: they were jawless.

How would being jawless affect an organism's life?

Out of the murk, with sinuous grace, the predator came.

Caught in the net, the haddock was unable to escape.

The predator sniffed, then touched, then clamped its circular mouth to the struggling haddock's side. Teeth set in the cartilage ringing the jawless mouth rasped into the haddock's skin and flesh.

Not even the most violent, desperate struggles of the restrained fish could deter the gnawing predator. Eventually the struggles diminished, then stopped—dead.

By the time the net was hauled aboard the dragger *Jaqueline*, the predator had burrowed completely into the dead fish and was feeding leisurely on the victim's intestines.

When rough handling by a deckhand burst the haddock's side, the wrinkle-faced predator spilled onto the deck oozing slime.

What was this revolting predator?

The crew of the dragger most often called them slime eels, but most textbooks call them hagfish. Call them what you will; they are wrinkle-faced, slimy, jawless fish.

Supplemental Activities

1. Find pictures and descriptions of ostracoderms for comparison with modern jawless fishes.
2. The story of the lamprey, another jawless fish, and its effects on the Great Lakes trout fisheries is fascinating.
3. Construct a classification chart showing the relationship of the jawless fishes to the rest of the vertebrates.
4. Obtain a lamprey or hagfish from a biological supply house for firsthand study of the anatomy of these strange creatures.

Mammals

144. Island of Wild Horses

What is the most dangerous, most dreaded island in the North Atlantic shipping lanes?

One hundred miles southeast of Nova Scotia lies a mariner's nightmare: a low island of constantly shifting sand. It is the graveyard of many ships and their crews. It is Sable Island.

Grazing on the coarse dune grasses, drinking from the brackish ponds, and coursing the sand dunes is a population of wild horses as mysterious as the island upon which they live.

Perhaps they sprang from horses brought from the mainland to pull lifeboats and relieve humans of other burdensome work. More likely they came from the wrecks of one or more ships of earlier times.

Records of their existence date to the early 1700's and are as mixed with legend as are the stories of the wrecks, wreckers, heroes, rascals, and ghosts of Sable Island, the island of wild horses.

Supplemental Activities

1. For some good ghost stories, tales of shipwrecks and wreckers, turn to the books of Edward Rowe Snow, a man who wrote much about this misty, mysterious island.
2. Locate Sable Island on a map or chart of the Atlantic Ocean. Then consider this question: Why has Sable Island been such a threat to shipping?
3. Write a poem or story about the horses of Sable Island.
4. Speculate as to how a population of wild horses would impact on and be impacted by an island environment.
5. *Misty of Chincoteague,* a popular children's book, provides clues about the story of an island of wild horses off the coast of the United States.

145. Look-Alikes

What happens when two groups of mammals compete for the same environment?

Ninety-five percent of all living mammals belong to a great group called placentals. But it was not always so. Early in the Cenozoic Era, the placental mammals were on more nearly equal footing with their only serious competitors: the marsupials.

After the downfall of the dinosaurs, these two mammal groups spread out, adapting to fill in recently vacated niches with new species.

Desert environments were open to small seed-eaters. There, the jerboa mouse (a marsupial) and the gerbil (a placental) arose. The opportunities for a small, fierce predator spawned Tasmanian devils and wolverines. On the open plains, kangaroos and antelope vied for supremacy.

The placentals—gerbils, wolverines, antelope, and their varied kin—won. Marsupial look-alikes survive only where they are not in competition with placentals.

Supplemental Activities

1. Extend the list of look-alikes by finding a book on marsupials, selecting interesting species, then trying to find a placental that occupies a similar niche.
2. How does it happen that Australia is so blessed with marsupials?
3. The only North American marsupial is the opossum. How has the opossum managed to survive in a continent dominated by placental mammals?
4. Distinguish between the two important terms scientists use in discussing the subject matter of this opener: parallel evolution and adaptive radiation.
5. Another ancient mammal group is now represented by only a few living species. Look up *monotremes* to begin learning about this group.

146. Rat Wars

What mammal (other than Homo sapiens) *is most dangerous to humans on the worldwide scale?*

No strangers to war themselves, the people of Europe were interested spectators of the rat wars of medieval times.

It is sometimes hard to sort fact from folklore in stories that are legends. However, there seems little doubt that the black rats of Europe were successfully challenged by the larger Norway rats that appeared to have immigrated from the Arab, not Scandinavian, countries in the Middle Ages.

Contemporary accounts speak of rat armies, led by wise generals and brave lieutenants who fought with cunning strategy and bold tactics.

The outcome of the rat wars would forever alter the course of human affairs, for the combatants brought destruction to humans.

Supplemental Activities

1. What famous ballet features a rat general commanding an army of rats—a legacy of the rat wars?
2. Which rat carried the black death through Europe?
3. Are accounts of rat wars available? Ask a librarian.

---

147. The Mammal That Won the West

What mammal contributed to the exploration, settlement, and development of the western United States and large parts of western Canada?

Any movie buff will tell you that John Wayne, James Stewart, and Gregory Peck won the West. Gun fanciers claim that the winning of the West should be attributed to Winchester rifles and Colt revolvers.

A strong case can be made for the beaver as the mammal that won the West.

In those early days of our country, beaver hats were all the rage in Europe. Trappers could do quite well for themselves supplying pelts for the European hatters.

Beaver trappers pressed ever westward in search of beaver and wilder and wilder lands to suit their dispositions. Settlements gradually followed in their wake.

Supplemental Activities

1. Several other mammals figured prominently in the development of the western states and provinces: horses, cattle, bison, and wolves. Each has a fascinating story awaiting the library researcher.
2. Write an essay about trapping in today's world.
3. Beaver are important conservation tools in much of the western parts of North America. Find out more about this.

148. Break a Leg!

For whom is "Break a leg!" an expression of encouragement and a way of saying "Good luck"?

"Break a leg!" traditionally means "Good luck!" for actors. Some experts in such matters suggest that it may refer to fancy bows of the sort that follow a well-received performance. Another somewhat more puzzling luck symbol is associated with acting: the rabbit's foot!

Both luck tokens may be connected historically.

Powdering your face with the left foot of a rabbit was a theatrical custom—one far more frequently practiced in past times than today.

The connection between breaking legs, rabbits' feet, and the stage is a tenuous one at best, but one originating in biology.

European rabbits are born with their eyes open, which, according to folklore, gives them the ability to see and stare down the Evil One. Acknowledging the obvious difficulties of applying makeup with rabbits' feet, we can only speculate on the twists that transferred the "luck" from one end of the rabbit to the other and then to actors—traditionally a superstitious lot.

Supplemental Activities

1. Research stories behind other animals connected with luck, both good and bad.
2. Find out what differences separate rabbits from hares. Then decide which category would be appropriate for Bugs Bunny.

---

149. Teeth: Black and Poisonous

Titian R. Peale stood on the edge of discovery as he carefully peered into a pitfall intended to trap a wolf. Peale, a famous wildlife artist of the last century, hoped to have a specimen to paint. And he did—only one far smaller than the wolf he anticipated.

What mammal has black, poisonous teeth?

At the bottom of the pit scampering frantically about was the smallest mammal Peale had ever seen. Smaller even than the mice that plagued Peale's camp, this little grey creature was the first short-tailed shrew ever seen by non-Indians.

The shrew had teeth with black tips and a poison powerful enough to subdue a beetle in seconds! This may not seem like a lot to you, but if beetles weighed almost as much as you, having a poison might be quite a comfort!

Supplemental Activities

1. Titian R. Peale also painted portraits. Find out which American president posed for Peale.
2. Shrews are abundant in your neighborhood, in all probability. House cats are a more likely source than wolf traps, but you can probably get a shrew to study. But be careful of their bite!
3. Most mammal guides will sketch in some interesting details about the natural history of these tiny poisonous mammals.

150. Puzzlingly Healthy Arteries

Which mammals don't need to worry about hardening of the arteries?

By all we know of the chemistry of lipids (fats), seals and sea lions should suffer from hardening of the arteries. However, they are not troubled by this circulatory disease, despite the fat content of their diets and their generally rotund figures.

Seals and sea lions do not produce apoprotein E, which, together with apoprotein B-100, appears necessary for metabolizing artery-clogging fatty deposits in other mammals, including humans.

How seals and sea lions get along without this chemical is a mystery. The fact is they have puzzlingly healthy arteries.

Supplemental Activities

1. Respond to this statement: Secrets of seal health may heal humans.
2. Explain hardening of the arteries.
3. Define apoprotein and relate it to the chemistry of lipids.
4. Debate the efficacy of using mammals in medical research.

---

151. Smokey's Ally

Which mammal prevents more forest fires than any bear?

Smokey the Bear is a widely recognized symbol of human efforts to prevent forest fires, yet another mammal has proved far more practical as an assistant in our effort to protect forests: the beaver.

For more than half a century, conservationists have been live-trapping beaver, especially from suburban areas where their tree-downing and dam-building habits make them few friends, and have been transporting them to remote areas.

Back in the hills the beavers quickly reorient themselves and begin damming the nearest water source. The ponds they create put a ready supply of water at the disposal of fire fighters in areas too remote for trucking in water, let along tapping hydrants.

Supplemental Activities

1. Speculate as to other effects of beaver, good and bad, and then research to see how well your speculations stand up.
2. Beaver actually modify their environment. Try to find other examples of animals that change geography.
3. Look into the story of the selection of Smokey, Woodsy, and other animals symbolizing our efforts to care for our environment.
4. Plan and role-play a strategy to make the beaver the symbol of conservation efforts for your province or state.

152. Terminal Velocity and the Cat

Do cats always land on their feet?

Although veterinarians, who do not wish to set bones broken unnecessarily, caution against tossing cats about, cats do have a remarkable ability to land on their feet after falling.

Recent studies have revealed some surprises: short falls often break legs and falls from upper-story windows often kill cats, but falls from greater heights still have a surprisingly large number of survivors.

A falling cat reaches a terminal velocity around a hundred feet into the fall. At that point the cat's legs extend and spread-eagle, causing the animal to land ventral surface first— spreading the force of the landing widely.

Supplemental Activities

PLEASE EXCLUDE CAT-DROPPING FROM ANY CONSIDERATION OF ACTIVITIES!

1. Discuss the reasons for restricting studies of this sort of accident to statistics rather than experiments.
2. When do humans in free fall reach terminal velocity, and what is our terminal velocity?
3. Impacts for humans at terminal velocity tend to be terminal! Why are humans not so lucky as cats?
4. Define "terminal velocity."
5. Investigate the rotational dynamics of the cat's ability to turn while falling.

153. The Adaptable Rat

Which wild animal is best adapted to humans?

Adaptation—change in structure or behavior to suit environment—is one of the major concepts of biology. Few better examples can be found than the common, everyday rat.

Although rats are rodents, gnawing mammals basically suited to a plant diet, rats will eat all foods you eat and many things you would not even consider: paint, leather, even the toenails of living elephants!

City rats gnaw through pipes for drinking water. They climb brick walls, gnaw into cinder blocks, and even survive being flushed down toilets in tenements.

Country rats gnaw through the insulation of electric wires for reasons not well understood but with shocking and arsonous results.

Rats' greatest adaptive triumph is developing poison-resistant strains, thwarting our efforts to control their populations.

Rats even survived nuclear weapons tests on Pacific islands!

Supplemental Activities

1. Find out what your community does to control rats.
2. Obtain a rat skull and study the teeth. Find out how and why rats must constantly gnaw to tend their teeth.
3. Read studies of experiments involving rat intelligence and behavior.
4. Why are rats and mice used as substitutes for humans in medical research?

---

154. Double-crossing Glands

How many ways can your skin double-cross you just before a date?

Humans, like other mammals, have glandular skin. Yet for humans, these glands seem to let us down or even double-cross us.

Sweat glands are especially important both in cooling the body and in removing wastes. Unfortunately, especially after puberty, human perspiration becomes noticeably malodorous—a fact that can have a negative impact on social relationships.

Sebaceous glands lubricate mammal hair on fur, making it lustrous and water-resistant. Yet for the "naked ape" (as Desmond Morris described us), sebaceous glands on the face can and often do become unsightly pimples—especially just before a big date.

Scent glands help many mammals attract mates or mark their territories. Ours—for better or for worse—do neither. In fact, they are vestigial. Woe to us and our double-crossing glands!

Supplemental Activities

1. Although offered facetiously, the opener does make some points worthy of serious consideration. Rate the three kinds of skin glands in order of their apparent usefulness to humans, then check your rating by researching each.
2. Mammary glands, functional only in the breasts of females, provide milk for all species of young mammals. Investigate the evolution of these glands and report your findings to the class.

155. Humankind's Worst Enemy

Which mammal has killed the most humans, destroyed the most of their property and goods, and spread the most disease?

The answer is not humans!

The worst enemy of humans is the most numerous and widespread of all mammals on our planet: the rat.

Rats destroy one fifth of the world's food supply, ironically doing the most damage in Third World countries, which have the least food to spare. The Indian rice destroyed by rats could feed 250 million people a year.

The number of people bitten by rats is in nine figures.

Rats spread several of our most dreaded diseases, including bubonic plague, which killed a third of the people in medieval Europe.

Not all of our wars have done so much damage to us as have rats. Surely rats are people's worst enemy.

Supplemental Activities

1. The old saying is "Dog is man's best friend." In view of this opener, suggest an update that goes beyond eliminating sexist language.
2. If rats are so bad, why don't we just trap or poison them all? Research into this question will surprise you.
3. Not to your taste perhaps, but many of the people who will eat meat today will eat rats. Find out how important rats were to the Polynesians as they settled the Pacific islands and what proportion of the diets of Third World people consists of rat.
4. Considering all the odd and specialized cookbooks published, can you find a rat cookbook?

156. The Home Turf Advantage

Is there really an advantage to playing on the home turf?

Whether or not it matters in sports, it does matter in the lives of many mammals.

The raccoon was methodically, if none too neatly, sorting the author's garbage. Edibles were being eaten, inedibles scattered all around the garbage can.

With only passing interest, the small black cat had been watching this vignette. However, when the author came around to see what all the sorting and munching noises signified, the cat tensed.

The cat's tail hairs stood out like bottle-brush bristles. The cat snarled, then lunged at the raccoon.

The raccoon, even though he was three times the cat's size, far heavier, and probably far more experienced in combat, took off at a dead run—never even seeing the author.

The cat pursued only to the edge of its yard.

Such scenes are enacted often in suburban backyards and remote wilderness places. A smaller animal chases off a larger one—when the smaller animal is defending its home turf.

Supplemental Activities

1. Look up *territory, territoriality,* and related key words to see what the literature says about the home turf advantage.
2. Peter Benchley, the author of *Jaws,* based his story of multiple attacks by the same shark on research into territorial behavior. You can find out more about this theory in any of several currently popular books on sharks.
3. How might recognition and respect for territories be an advantage for both the aggressor and the defender in nature?
4. Why might the cat have gone after the raccoon when the cat's owner came on the scene?

Mollusks

157. Boy Defeats Giant!

Of all the legendary sea monsters—sea serpents, giant squid, man-eating clams, and huge octopuses—which really exist?

"BOY DEFEATS GIANT . . ." the headline read.

This story, a variation on the man-bites-dog theme, is not a retelling of David and Goliath.

The boy, Tom Squires, his father Daniel, and another man, Theo Piccot, set out of Conception Bay, Newfoundland, for a day of squid jigging.

Their dory bumped a sleeping giant squid, which immediately wrapped tentacles around the dory and became decidedly grumpy about their little accident.

The two men were petrified with fear. Twelve-year-old Tom, however, decided to sell his life dearly. He grabbed a heavy fish knife and hacked off part of a tentacle.

The squid released the dory.

The men, suddenly mobilized, rowed for shore more spiritedly than they had set out.

The section of the tentacle left writhing in the boat was reliably measured at 19 feet 6 inches!

Supplemental Activities

1. Using a real squid or a clear picture, estimate the length of a squid that could have a tentacle 19.5 feet long.
2. Giant squid are regularly caught by fishing boats, especially Japanese boats. A little library time should yield accurate dimensions for your normal, everyday giant squid.

158. Of Snails, Slaves, and Traders

How might shells help archaeologists reconstruct trade routes of early humans?

Lewis and Clark came into possession of the dress of a Cree woman. The details of the story are not too clear, but the decorations upon that dress pose even more of a mystery. Forty-eight cowries, highly polished half-inch-long snails, adorned the garment. These particular snails came from the Indian Ocean—half a world away from the Cree nation.

How those snails came into the hands of Plains Indians is a story of trade routes, goods, and contacts among humans that remain a mystery.

This kind of cowrie had long been used as money, for buying and selling commodities including slaves. The practice has been so widespread, the shells are known to this day as money cowries.

Supplemental Activities

1. Shells, because of their durability, are especially useful to archaeologists attempting to rebuild lost trade routes. Money cowries found in the Indian mounds of the midwestern United States have been traced to trade goods brought to the New World by Christopher Columbus. Consult archaeology books for more such connections.

---

159. The Badge of Christian Soldiers

How many symbols can you name that involve shells?

The Crusaders marched along the shore. The hot Mediterranean sun glinted on their armor, sparkled on their perspiration, and glowed warmly on the scallop shells affixed to their helmets. Most of the men had collected these shells as marks of their participation in the efforts of the English to free the Holy Land.

Scallop shells, especially shells of *Pecten jacobaeus,* were also proudly worn by pilgrims to the tomb of St. James the Apostle at Santiago de Compostela, Spain.

Considering the long history of scallop shells as Christian symbols, it is not surprising that many English coats of arms bear representations of these shells—badges of Christian soldiers.

Supplemental Activities

1. Survey books on art for other representations of scallops.

2. Which goddess of the Greeks, then the Romans, is often depicted stepping from a huge scallop shell? What is the story behind that scene?

3. Find scallops in a book of heraldry. You may discover, whether or not you are English, that your own family has a coat of arms.

4. Coffin handles, tomb decorations, and gravestones often have scallop designs incorporated into them. Research and speculate on the symbolism of this use of the scallop, perhaps even the pagan connections involved.

160. Royal Purple

How much trouble would it be, how much money would it take, to turn your white tee shirt into a purple tee shirt?

For ten days and nights the Phoenician slaves stirred and skimmed the great copper kettles of foul-smelling broth and toiled to keep the fires going. Inspired by the whip, guided by skilled masters, the slaves contributed to wealth they would never share.

Days before the boiling began, other slaves had collected several kinds of living snails—among them *Murex brandaris*—crushed their shells, picked out the meat, salted it, and allowed it to cure.

Secretions from a tiny gland in each mollusk began changing color upon exposure to light. The curing and cooking would concentrate and darken these secretions. At length, the masters would soak cloth for as much as a day to achieve just the right shade of purple—maroon to our eyes.

Beneath sails of such purple cloth, Cleopatra would entertain on her royal barge. Purple cloth would grace the Tabernacle, the wrappings of Egyptian mummies, and the robes of Rome's senators. Ever after, purple would be associated with royalty and nobility.

Supplemental Activities

1. If you have not yet studied the Phoenicians, look them up in an encyclopedia or ancient history text. Their exploits in business and seafaring make excellent reading.
2. Making the dye was one of the most closely guarded industrial secrets of ancient times. Other interesting secrets include the embalming techniques of the Egyptians, headshrinking methods of certain tribes, and the formula for Coca-Cola! You can find books that will let you in on some of these great secrets; others remain mysterious. Industrial espionage is a realm of human endeavor little noted in movies but very real and probably more lucrative and somewhat less dangerous than international spying. It would certainly make for entertaining reading.

161. Honk a Conch

How do you play a shell?

Sonorous and mournful, the note carried over the distant reef surf, the lowing of the palm fronds, and the sobs of the people. The note carried the people's grief to their gods and announced the death of a Polynesian warrior chief.

A note sounded by human breath resonating in the spirals of a great sea snail seems a most appropriate way to honor a man of the sea.

Students on their way to classes at Harvard University in its earliest days were urged along by a student trumpeting the same note on the same kind of shell: Triton's trumpet. The periods of the Harvard students' school day were sounded by a honk on a conch.

Supplemental Activities

1. Triton's trumpet was the shell used by the islanders and the Harvard timekeeper, but any large conch shell can become a trumpet. To make your own, saw or file off the tip of the spiral of a large snail. Make a trumpeter's lip and blow! Your first attempts may sound more comic than sonorous, but with practice, you should be able to make decent honks on your own conch.
2. Read *Lord of the Flies,* giving special attention to the shell and its significance to Ralph, Roger, and Piggy. Then discuss this with the class.
3. Determine what you hear when you press a large shell to your ear. Is it "the sea"?
4. Delve into the physics of shell sounds to explain and compare the "sounds of the sea" when holding a large shell to your ear and the trumpeting sounds when you blow through such a shell.
5. Make trumpets of several different shells varying in size and species. Predict what sounds they will make, then experiment to test your hypotheses.
6. Experiment with shell music. Can more than one note be sounded on a conch? Can additional holes make shells into sea flutes? Can shells be made musical in other ways?
7. Might rock music give way to shell music?

162. History Collectors, Shell Collectors

What did the insane emperor Caligula have in common with hundreds of normal, sane people in our country?

Poor, mad Caligula and many normal people collect shells and the bits of history, both natural and human, they embody.

Archaeologists find shells in ancient graves and ruins, often shells from places far distant from their dig sites. Such finds speak of humans' long history of love for shells, the flesh of the mollusks, and routes of trade in forgotten times.

Caligula aborted an invasion of England in A.D. 40 to have his legions collect shells. By the time of his departure for Rome, Caligula had a very large collection of the shells of the English Channel area and a very puzzled army.

Since Caligula's rule, the who's who of shell collecting has included many names stressed in history classes. Collecting shells is collecting history.

Supplemental Activities

1. Admiral Chester Nimitz was a famous military leader, far more highly regarded than Caligula, who also encouraged shell collecting. Check out this lead for an interesting sidelight on World War II.
2. Captain William Bligh brought shells home to his wife. If the name only sounds vaguely familiar, look Bligh up for another great story and a bit of the history of shell collecting.
3. Interview a shell collector, or arrange for the collector to speak to your class.

163. St. Augustine Monster

How big are the biggest invertebrates? How big an animal might still live undiscovered beneath the sea?

A great smelly mass of flesh washed ashore in St. Augustine, Florida, in 1897. About that, there can be no doubt.

Reports of its size vary and are complicated by the fact that a substantial portion of the mass remained buried in sand. It was too large to be dug out of the soft sand and too heavy to be dragged.

Samples were salvaged and are preserved to this day in the Smithsonian Institution.

The best experts then and now are unsure what the decayed mass was, but the most often advanced theory is an octopus—part of an octopus, that is. If the mass was part of an octopus, the live octopus would have spread out over a 200-foot-wide circle as it searched the sea floor for food!

Supplemental Activities

1. Research librarians may be able to produce information on this monster if you'd like to form your own opinion.
2. Monster stories abound. In all probability, your area has some. Investigate your local monster story. Report it to your class for critical evaluation.
3. Can you find out what psychologists think makes monster stories so interesting to us?
4. Write a story, play, or skit about monster busters, professional monster-story researchers.

---

164. State Mollusk Debate

In schools, homes, and the halls of government throughout California, there is an ongoing debate. Will the state mollusk be the abalone, a commercially important food, or the banana slug, an oddly beautiful creature?

Although this issue will probably be resolved, it raises a more far-reaching question:

What purpose is served by declaring a state flower, stone, mammal, fossil, or bird?

Supplemental Activities

1. Discuss the value of making biological things state symbols.
2. Does your state or province have official symbols?
3. It has been suggested that promoting a state "whatever" is a good first experience in politics for students. If you are interested in politics, consider starting work toward such legislation. Few provinces or states have official slime molds, worms, jellyfish, or parasites.

165. The Parasite That Wasn't

Give some examples of mistakes scientists have made.

Cuvier, the famous French zoologist, named the parasite. Von Kolliker described the worm's circulatory, digestive, and respiratory systems in detail.

However, the worm that Cuvier discovered and von Kolliker described was not a parasite, not a worm at all, not even an unwelcomed addition to the blanket octopus in which it was found.

The slender tubular thing with suckers on its "head" was the detached copulatory arm of an amorous male octopus.

When blanket octopuses mate, the male's spermatophores—packets of sperm—are passed to the female by a special arm which reaches up, under her mantle, then breaks off to remain with her for some time.

Supplemental Activities

1. Scientists make mistakes. Aristotle thought eels were horsehairs come to life. Hitchcock mistook dinosaur tracks for bird tracks. These and the mistakes made by Lamarck and others make entertaining reading which will also give you insight into what it means to be a scientist.
2. The reproductive behaviors of other cephalopods are worth looking into.

---

166. Holy Migrating Mollusks, Batman!

Is migration the sole province of birds?

Several years ago, the comic strip "B.C." let out the secret: "Clams got feet!" Biologists, of course, have always known that mollusks have feet.

Migrating clams are a pheneomenon of sandy shores and several species scattered around the world.

The little clams *Donax,* sometimes called coquinas, migrate up the beach with the incoming tide and down the beach as the tide ebbs.

They literally jump out of the sand as large waves surge up the beach, then dig in before the backwash can negate their progress. They use their foot like a sea anchor to keep from being swept too far up or down the beach as they migrate.

Supplemental Activities

1. All behavior is purposeful. What do you think this migratory behavior might accomplish? Hint: Ask yourself what clams might want to obtain or avoid.
2. Other mollusks—limpets, for example—migrate, or at least travel. You might enjoy studying their actions.
3. Wherever you live, some kinds of mollusks live. You might be able to set up an experiment or observation program to see if local mollusks move with regularity and purpose.

Moss

167. Alien Invasion

Thousands of the little sealed chambers had been released from the pod to drift aimlessly through the void. Some settled in vast bodies of water, eventually to sink and drown; others died dessicated on barren land under a merciless sun.

But a particular chamber landed on a damp, shady, rocky outcropping. The being within sensed favorable conditions. The outer shell of the life support system parted. A slender green life form emerged to grow and change.

What is the invading organism?

In the years to come, the alien and its descendants would cover the ledge with a lush, green miniature jungle.

The sealed chamber was a single moss spore, the little being a protonema.

In this way moss spread, making green and habitable rock, bark, and other surfaces.

Supplemental Activities

1. Raise moss from spores. Several biological supply houses offer kits requiring only water, sunlight, and some patience.
2. Examine prepared slides of moss spores and protonemas.
3. Hunt for moss protonemas with a powerful hand lens.
4. Research and discuss alternation of generations in terms of plant evolution.
5. Explain why so many lower forms of life produce vast numbers of reproductive structures: spores, eggs, seeds.

168. Third World Fuel

What fuel grows in northern wetlands?

Over thousands of years, in many parts of the world, depressions from glaciers or other forces fill with water from rain or springs. Plants overgrow the shallow edges. The dead plants are unable to decay completely in the stagnant, oxygen-poor water, which becomes increasingly acidic. Sphagnum and other acid-tolerant plants dominate, and a bog forms as layers of partially decayed material accumulate below.

What fuel is forming in these bogs?

We call this layered, partially decayed material peat.

The peat, cut out in blocks or pumped out as organic slurry and then dried, forms potential fuel. In Ireland, Denmark, and other northern and Scandinavian countries, peat has kept the home fires burning.

Now Africa, Indonesia, and other parts of the Third World are looking to bogs as a source of cheap fuel to propel them into an industrial future.

Supplemental Activities

1. Bogs which take thousands of years to develop can be depleted in tens of years. What are the implications of this fact for industrial development, ecology, agriculture, economics?
2. Are bogs part of your province's or state's economy? How are they being managed, developed or protected?
3. Bogs support unique ecosystems. What factors should be considered before sacrificing them and the species they contain for industrial development?

---

169. Most Useful Moss

How many uses for moss can you come up with?

Under the microscope the cells explain the many uses of sphagnum moss. Tiny living cells and large, empty, warehouselike cells alternate in the thin leaves. The empty cells soak up water like a sponge.

Sphagnum moss has diapered Indian babies. Quite possibly sphagnum provided the first disposable diapers, new ones of dried sphagnum being available in any bog.

Desperately short of medical supplies in the trenches during World War I, German doctors packed wounds with sphagnum moss and rediscovered the healing power of moss, touted in folklore.

Buried with fire-heated rocks and fish or meat, wet sphagnum moss turns a hole into a slow cooker for the backpacker.

Sphagnum is a most useful moss!

Supplemental Activities

1. These and other uses depend upon the absorbent or water-holding power of those enlarged dead cells. Speculate as to how the living plant would use these cells.
2. Sphagnum moss is added to garden soil even though it is not especially rich in nutrients. What function might sphagnum serve in the garden?

170. The Bog Murders

Summoned to the misty bog at Tollund Mose, Denmark, Dr. Glob stared down at the human foot and shoulder, naked and discolored by the dark peat that had covered them.

Ever so carefully, Dr. Glob and the policemen dug out the whole body: a man, naked but for a belt, a skull cap, and the leather thong that had choked the life out of him.

What does this evidence suggest to you?

Dr. Glob was excited by the challenge of this strange case. Not a forensic expert, but an archaeologist, he knew what the police only suspected: the murdered man had been in the bog a very long time. The man was not the victim of some weird cult but an Iron Age man sacrificed to long-forgotten gods some 2,000 years earlier!

From this amazingly well-preserved body—skin still soft, stomach contents complete—Glob and others have been piecing together the story of these bog deaths.

Supplemental Activities

1. Although you are most unlikely to find human remains, a little digging in a local bog might unearth some interesting specimens.
2. Human remains have been found preserved in bogs in the southern United States. Their discoveries make for interesting reading.
3. Volcanoes, floods, and other natural disasters have preserved people and prehistoric organisms. The sciences of archaeology and paleontology are tools to put together the jigsaw-puzzle pieces of the past. You might like to read about these sciences for general interest, or perhaps to line up interesting college course choices or even career choices.

Plankton

171. Invisible Animals

What advantages does invisibility offer plankton?

Most plankton have limited mobility at best. Few can swim as actively as the animals that feed upon them.

Some plankton are transparent—for all practical purposes, invisible—a good defense against visual predators.

Supplemental Activities

1. Why would invisible plants be less likely to exist than invisible animals?
2. Justify this statement: Transparency would not save plankton animals from their largest predators.
3. Are there any transparent animals on land?

Plant Life

172. Solar-Powered Food Factories

When was solar energy first used?

As the prices, monetary and ecological, of energy have gone up, we have looked with increasing interest at a free, nonpolluting energy source: sunlight.

Although only a small part of the sun's energy reaches the earth's surface, it is an energy source worthy of consideration.

Solar power is relatively new to us but biologically ancient. For billions of years plants have used solar power to make food.

Leaves are solar-powered food factories. These plant organs are composed of tissues that bring raw materials—water, minerals, and carbon dioxide—together in the presence of a catalyst to make food with the energy of sunlight. The process: photosynthesis.

Supplemental Activities

1. What do we call the process? the catalyst?
2. Study the various solar power units available for home use and compare them with leaf tissues. How are they analogous? different?
3. What are catalysts? What is their advantage in biological systems over analogous systems in technology?

Protists

173. Microspeeders

How fast can microscopic organisms move?

Rolling, rotating, yawing, and pitching like some space vehicle run amok, the tiny protists move; it is nearly impossible to view them under the microscope.

Paramecia can hit speeds of 2 mm/second! Although it doesn't sound very fast, this amounts to moving seven times their body length per second.

To view these microspeeders, it is usually necessary to drug their water.

Only when paramecia are slowed down can their motion be studied and appreciated. The cilia covering their tiny bodies beat with the rhythm of grasses in a windswept field.

Supplemental Activities

1. Calculate the speed of paramecia from the information given and express your answer in kilometers or miles per hour.
2. What chemicals can you use to drug protists?
3. Experiment with cooling to see if you can slow protists without exposing them to drugs.

174. Scourge of the Tropics

Most of the men were sick. They would burn with fever, then shiver with chills despite the oppressive heat of the jungle. Then, for a day or two, they would seem to improve. But the fever always returned, and the men worsened.

The captain turned to the ship's doctor. He had to know what was wrong and what to do. Their expedition some 300 years ago into the Central American jungles was their chance for riches and glory.

The doctor shook his head. "It is the air, my captain; the men have the bad air fever!"

Do you know what was really wrong with these would-be conquistadors?

It has taken us hundreds of years to know the fever comes from tiny protists, not "bad air," although we still call the disease *malaria.*

Supplemental Activities

1. Learn the life cycle of *Plasmodium vivax,* the malaria-causing organism. Then suggest strategies for protecting humans from this scourge of the tropics. After you have made your decisions, check to see what is actually being done.
2. What was there about early medical beliefs that made the possibility of bad air causing disease seem very reasonable to the doctor and captain described above?
3. Cholera was believed to be a result of breathing bad air. A bit of research into cholera epidemics of London will yield a story of a statistician's contribution to an understanding and eventual control of a terrible disease.

Protochordates

175. Disposable Houses

We live in a time when paper, plastic, and other materials encourage us to dispose of utensils, containers, plates, and many other things. We have yet to come to disposable houses, but they may be on the way.

What animal does live in disposable houses?

Nanoplankton are made of the smallest of all planktonic plants. They are tiny even by plankton standards. To feed successfully upon them, very special adaptations are required.

The disposable houses of appendicularians are such an adaptation.

Appendicularians are small relatives of the tunicates. An appendicularian looks somewhat like a bent tadpole with its tail attached to or towards the middle rather than the end of its body. Appendicularians make elaborate, nearly transparent homes of mucus. These houses have openings to allow water in and out through a system of filters, screens, and traps. Appendicularians beat their tails rhythmically inside a central passage to create a continuous flow of water through their houses. Objects too large to be ingested are blocked by the screens. Nanoplankton, up to a quarter of a million cells a day, are concentrated and eaten by the appendicularians within the mucus maze.

As often as six times a day, the mucus houses become contaminated with wastes and are discarded, with new ones built in their place.

Supplemental Activities

1. The discarded houses of mucus remain intact and adrift for some time. Speculate as to the impact of these on plankton on the surfaces of a floating world, as a source of organic material, and as a potential trap.
2. Research what is suspected about the role of appendicularians in chordate evolution.

176. Growing Down

Here is a science term so filled with familiar word parts, you may be able to define it if you just think about it for a moment or two: retrograde metamorphosis.

What is retrograde metamorphosis?

For a few hours of their "childhood," tunicates seem bright and full of promise. They are built like little tadpoles. They have the three distinguishing characteristics of chordates: paired gill slits, a dorsal nerve cord, and a notochord. These actively swimming little creatures seem well on their way to growing into competent little fish or something similar.

However, nothing of that sort happens!

The tunicates attach themselves to rocks or other solid surfaces, assimilate their tails, muscles, most of their dorsal nerve cord, and notochord, and turn themselves end for end in a toughening jacket of cellulose. For the rest of their lives, tunicates are sessile filter-feeders, little more than living lumps pumping water in one opening and out another. They are an example of retrograde metamorphosis, "growing down."

Supplemental Activities

1. Set up a chart comparing three kinds of metamorphosis: complete, incomplete, and retrograde. Try to find several examples of each.
2. What significance might retrograde metamorphosis have in the evolution of chordates?
3. Do tunicates reproduce before or after retrograde metamorphosis?
4. What is unusual about the tunicate ability to make a coating, or tunic, of cellulose?

Reptiles

177. The Star That Almost Fell

What would it be like to be bitten by a poisonous snake? What would you feel physically? emotionally? What would you do?

What one person did is the story of a star who almost fell.

About mid-morning on December 31, 1928, the curator of reptiles at the Saint Louis Zoological Gardens and his assistant began treating a Gaboon viper for parasites. It was as routine a procedure as possible with a venomous snake—routine until a fang grazed the curator's finger!

Despite the immediate, excruciating pain, the curator made an incision and began sucking poison from the wound. He started for his office, only 20 yards away, but became so dizzy he required help to get to a chair.

He received his first dose of antivenom within 15 minutes of the bite. He was in a hospital 15 minutes later, but experiencing terrible pain and great difficulty in breathing.

Before his release from the hospital 20 days later, he was to experience an agonizing variety of symptoms and one near brush with death.

His recovery was complete. Almost exactly 40 years later, he began a career as one of television's most respected spokespersons for wildlife. The young curator was to become the host of Mutual of Omaha's *Wild Kingdom:* the late Marlin Perkins.

Supplemental Activities

1. View episodes of *Wild Kingdom* in syndication or on tape if possible.
2. Locate pictures and background on the Gaboon viper, regarded by Raymond Ditmars, an expert on snakes, to be "the world's most frightful-looking snake."
3. Interview, in person or by telephone, an expert on snakes concering hazards and precautions that are part of handling poisonous snakes.
4. Especially if poisonous snakes are found in your area, learn how to recognize them and what to do if a person is bitten by one.

178. How's That Again?

The boy attacked the alligator!

That's right. The boy, a rather small 11-year-old, attacked a mature alligator, bare-handed!

Under what circumstances, if any, would you attack an alligator?

David Peters saw the alligator grab his dog. Without really giving the matter the thought it deserved, David lunged at the alligator, who was twice the boy's length and a whole lot stronger.

The alligator, not one to let such things pass, bit the boy's right arm.

As it happened, David had more friends on shore than the alligator had in the water. Several of David's friends pulled him free of the alligator.

The pup got quickly ashore. David healed. So, unless you were rooting for the reptile, the story has a happy ending—better than people who attack alligators have any right to expect!

Supplemental Activities

1. Unfortunately, several people-meet-alligator stories have ended in human deaths. Investigate some of these stories and develop some reasonable safety guidelines for people who live in 'gator country.
2. How do such stories affect conservation efforts aimed at preserving the dwindling numbers of the crocodilian reptiles?
3. Role-play the parts of a Florida land developer, state conservation officer, and concerned citizen to develop a feel for the issues around the interaction between humans and alligators.

179. Smarter Than the Average Reptile

They chirp, click, hiss, growl, grunt, and bellow at each other and apparently convey some degree of meaning with their sounds. Underwater they emit sound signals inaudible to our ears but clear to theirs.

They gesture to each other by raising their snouts above the water, arching their backs, blowing bubbles, and rushing.

Some species have been observed hunting cooperatively.

A feeding group will sometimes share food. Small prey, which flops about rather than breaking conveniently, will sometimes be held by one so another can rip off a mouthful.

Have you figured out which reptiles do these things?

These are crocodile traits. Because of such complex behavior, scientists believe crocodiles are smarter than the average reptile.

Supplemental Activities

1. Crocodiles are superior to the rest of the living reptiles in some anatomical ways. Find out what these are.
2. Some prehistoric crocodiles are so incredible, you ought to learn about them. Dryosaurus, orthosuchus, and the mesosuchians would be good ones to start with.

---

180. Turtle Comeback

What favorite pet for generations of children was removed from stores because young purchasers were dying to buy them?

Some children actually died shortly after buying these pets.

Since 1976, when the United States government acted to protect children from salmonellosis, it has been illegal to sell painted turtles for pets.

Recent studies at Louisiana State University have provided a means of chemically treating turtle eggs to kill the salmonella bacteria.

The pet turtle industry may be on the verge of a comeback.

Supplemental Activities

1. Respond to this premise: The rebirth of the pet turtle industry would be good news and bad news.
2. Salmonella bacteria *inside* hen eggs have been blamed for human illness and some deaths on the East Coast. Look into this far more serious issue and report your findings to the class.
3. Painted turtles, naturally quite colorful, were often painted before sale. How was this practice harmful to the turtles?

181. A Frog's Fate

What is a frog's worst nightmare?

"Time is fun when you are having flies" might have been the frog's thought, if frogs had thoughts. At the pond's edge, on a sultry summer day, the frog was contented—contented and doomed.

Hidden in the sedges, a nightmarish enemy was present. A forked tongue sampled the tastes and smells in the air. "Frog" registered in the snake's brain.

Slithering stealthily, silently, the snake homed in on the faint signals.

Without warning, from behind, the snake struck.

Startled, the frog tried to leap away, jamming the snake's back-pointing teeth deeper into its leg muscles. In the struggle, the snake threw coils around the frog, then clamped it in a crushing grip.

Head to head, the snake confronted the frantic frog.

The snake's jaws unhinged as she closed her mouth over the frog's snout. Again the frog was held from escape by back-slanted teeth.

As the snake began swallowing the frog, head-first, her skin stretched, isolating scales like shiny islands in a branching river of translucence. Her glottis, snorkel-like, extended from her lower jaw so she could breathe and swallow simultaneously. As the frog, struggling still but with decreasing strength, slid deeper into the snake, her esophagus stretched; her ribs, unrestricted by a breastbone, expanded.

Such a death is a frog's fate.

Supplemental Activities

1. Investigate the evolution of amphibians and reptiles and the number of species in both groups at various times. Then explain how this story is an allegory of the amphibian condition.
2. Read some more about hunting and feeding adaptations of snakes. Then write an essay entitled "Snakes, a Study in Specialization."
3. Why do snakes customarily swallow frogs head-first?

182. Locomotion Without Limbs

A few years ago a television short showed a brave child, born without a single complete limb, dress herself unassisted. She did what would seem impossible to the rest of us.

Moving without limbs, as snakes do, may also seem impossible until you think about it.

How do snakes manage to move without legs or arms?

With power and grace, but without limbs, snakes move.

They move by mastering a complexity of muscle, scale, and bone. Listen to and visualize this description of the structure of snakes.

Snakes have 200 to 400 vertebrae and muscles which connect: vertebrae to vertebrae, vertebrae to ribs, ribs to ribs, ribs to scales, and scales to scales. Tendons connect segments, both adjacent and distant, to other segments.

Coordinating these muscles, bones, and scales, snakes move straight by literally walking on their belly scales; in wavy curves, by pressing back loops on side loops; and even by throwing themselves along in S curves, sidewinder style. Thus snakes accomplish the seemingly impossible, motion without limbs.

Supplemental Activities

1. Compare the structural basis of the motion of snakes with that of eels, caecilians, and worms.
2. Find descriptions of snake motion in prose and poetry.
3. How many backbones, or vertebrae, do we humans have? How many are movable?
4. Capture and observe some **nonpoisonous** snakes to see for yourself motion without limbs.

183. They Do Have Funny Noses

List some animals or plants that no one would believe existed unless they actually saw them.

Gavials have funny noses. Well, sometimes they have funny noses, those slender-snouted crocodiles of India.

Gavials slash the water with their extremely slender jaws, frightening fish into frenzies. With a snap and a flip, a gavial will toss a frightened fish into the air, then catch and swallow it head-first. If you think about swallowing whole, live fish, taking them head-first is probably the smoothest way to do it.

Old woodcuts and reports from early explorers placed a bulbous mass on the very tip of the gavials' snouts. This feature was not taken seriously until recently, when it was established that some male gavials do have funny, bulbous noses!

Supplemental Activities

1. On the off chance that you might not know what gavials look like, find a picture of one. You will see an example of an animal no one would believe unless they saw one.
2. Just in case you are ever a contestant on a quiz show, learn the differences between alligators, crocodiles, and gavials.
3. Crocodiles have a bad reputation as people-eaters. For at least three kinds of crocodiles, this reputation is well-deserved. To locate stories, factual and not so factual (but no less interesting), look up alligators, Nile crocodiles, and Australian sea crocodiles.
4. After doing your research, you might write a response to this premise: Crocodiles have more to fear from people than people have to fear from crocodiles.

184. Gator Geography

After a bit of thought, respond to this statement: Only humans among all the animals can alter the geography of earth.

In arrogance and error, we once thought only humans significantly altered geography.

American alligators dredge out a home, or 'gator hole, enlarging and shaping it to suit their needs for cooling water and feeding. The spoil banks of these holes—mud, rocks, and organic materials brought up by the alligators—form little lush islands.

Female alligators push up mounds of mud, rocks, and vegetation for nesting sites—yet more little islands.

Upon these banks and mounds, plants—often different from surrounding water-loving species—establish themselves, creating variety in the environments and opportunities for any organisms that can share space with their reptilian neighbors.

Supplemental Activities

1. Come up with a list of other animals, with and without backbones, that reshape geography.
2. Research such geographic reformers as beaver, otters, and prairie dogs.
3. Tool use is another "human" skill we are now observing in lower animals. This would make a good research or discussion topic too.
4. Are there things only humans can do?
5. Alligators, beaver, and egrets share another connection with humans, a more dangerous one: fashion. Find out how human fashions nearly exterminated these and other animals.

Roundworms

185. Worms in Your Apples

The old joke is: What is worse than finding a worm in your apple? Finding half a worm in the apple you just bit into.

What could be worse than that?

Half a worm in the apple you just nibbled is bad enough, but imagine 90,000 worms in your apple!

Impossible?

Not at all. Tiny, threadlike roundworms are that abundant in soil, living animals, and plants including your lunchtime apple.

One scientist extracted 90,000 roundworms from an apple he had allowed to "mature" in a sealed chamber in his laboratory.

Roundworms are so abundant that a parasitologist once said, "If the earth and all of life except the nematodes vanished, the worms would outlive all!"

Supplemental Activities

1. Biology books can show you how to make a simple apparatus for collecting roundworms from soil.
2. Examine some living vinegar eels under a microscope.
3. Allow a wild apple to decay partially inside a closed jar. Then search for nematodes, roundworms.
4. Ask pet owners or a local veterinarian to provide samples of roundworms from dogs and cats.
5. Investigate the stories of some of our roundworm parasites—if you have the stomach for it.

186. Moose Sickness

How can you tell whether a wild animal you see is healthy?

Halfway across Main Street, in Bar Harbor, Maine, the moose stumbled. For a moment the people crowding the sidewalks thought the animal was falling down, but the moose recovered, staggered, then circled several times to the left as it drifted across the street.

Without seeming to notice the people scurrying out of its way, the moose stumbled onto the sidewalk. For a minute or two, the animal stared into the window of the Rexall drugstore; then, to the store owner's relief, the moose crossed the street.

By the time the game warden arrived, the moose was down on its knees in an alley beside the bank.

The warden and two police officers had a brief discussion.

"Moose sickness," the warden said, shaking his head.

The officers moved the crowds back.

The warden ended the moose's suffering with a single shot.

It would be 20 years before a Canadian researcher would demonstrate that the moose sickness was the result of parasites: brainworms.

Each year moose and some related animals face an agonizing death by paralysis, hunger, thirst, or possibly being torn apart by roving dogs: victims of the moose sickness.

Supplemental Activities

1. The lead question—How can you tell whether a wild animal you see is healthy—is especially important. Find out which animal diseases can be transmitted to people. This is a matter of personal health and safety.
2. Before 1965, no one knew the life cycle of brainworms. No one even knew that brainworms were the cause of the moose sickness. Research the life cycle of brainworms and suggest means of controlling the disease in wild animals.
3. The parasitic worms also attack wapiti, caribou, and deer. Deer seem to be the original host and are not much troubled by the disease. Theorize why this might be so.
4. Researchers and conservationists in Maine are presently attempting to restore woodland caribou to a wild area called Baxter State Park. How might they protect the caribou from possible attack by brainworms? How might the location, a state park, complicate the solutions you generate?

187. Killer of Americans

"It ain't nothin' but ground itch!" the poor woman assured her son.

The boy's bare feet began itching furiously shortly after he scampered back along the muddy path from the outhouse.

"It'll go 'way soon 'nough," she promised. "It ain't nothin'."

Well-meaning but ignorant, the boy's mother was very wrong about her boy's itchy feet.

What was the real problem?

This little vignette was once commonplace in the rural South of the United States and still is throughout much of Mexico, Central America, and northern South America. That boy and thousands of others were developing hookworms, a parasite so dangerous it was given the technical name *Necator americanus*, the "killer of Americans."

Infective larvae in the mud burrowed into tender skin between the child's toes. In his bloodstream, the worms would change and travel. Eventually the adult worms would clamp hooked jaws into the lining of the youngster's intestines and begin feeding—feeding on the boy's lymph, blood, and mucosa. In the end, the boy could die—another victim of hookworm—the killer of Americans.

Supplemental Activities

1. Examine prepared slides of hookworms with a microscope.
2. Speculate as to how children might be harmed by hookworms, then check out the effects in a medical book.
3. Trace the life cycle of hookworm, then identify ways to interrupt the cycle to protect humans from this dangerous parasite.

Segmented Worms

188. Anyone for Roast Worms?

What is the strangest thing you have ever eaten?

Just before dawn on the eve of the first day of the last quarter of the moon, the people of Fiji, Samoa, and nearby islands paddle canoes out to the reefs. Everyone is excited, anticipating an adventure and a feast.

Below the canoes, in shadowy crevices among the corals, the palolo worms writhe in anticipation of their own, unaware of the humans, the danger above.

Throughout the early autumn the worms have added new segments to their rear ends, their epitokes.

The segments of the epitokes contain almost nothing except reproductive tissue, sperm- or ova-producing cells.

The epitokes, almost spontaneously, break off and wriggle to the surface, shedding sperm and ova into the water. While the worms below return to their daily duties, the people above gather in the epitokes to roast at an annual feast.

Supplemental Activities

1. Anthropology books, *National Geographic* magazines, and other resources will provide you with information on other exotic foods.
2. Describe some things you eat daily that might not be palatable to other peoples of the earth.
3. Using biological clocks as a starting point, find other examples of events that are timed naturally.

189. No More Black Eye

How do you treat a black eye? How did your great-great-grandparents treat a black eye?

"Now see what you have done!" the woman yelled. "It will be blacked for sure and certain."

Like other mothers in the 1700's, this Virginia woman had been doctoring the family's minor medical problems—in this instance, a nine-year-old girl's tooth.

The woman had tried to pull it, the child had thrashed in discomfort, and one small hand had struck the soft tissue surrounding the mother's right eye.

"Now hush, child," the mother soothed. "It will be right soon enough."

As soon as the stubborn tooth was freed, the little girl was dispatched to the apothecary to buy a leech, "A big one for a pence!"

Later, the woman held the live leech to the discolored bruise until the worm took hold. Presently the discoloration receded as the worm filled itself with her blood. No more black eye!

Supplemental Activities

1. Explain how this folk remedy works by relating leech behavior to the cause of a black eye.
2. Find other uses for leeches in folk medicine.
3. Do modern religions that emphasize natural healing, such as Santeria, tout leech use?
4. If you live in a large city, see if you can buy live leeches from shops specializing in folk remedies.
5. Leeches have some other interest to medical researchers. See if a little of your own research can uncover the reasons for this.
6. Recently doctors have used leeches to relieve swelling in reattached fingers until veins are able to reestablish themselves. Find out more about this use of leeches in the high-tech business of modern surgery.

190. The Bite of the Painless Parasite

The girl paddled about in the warm shallows of the pond, unaware that she was being stalked. She would remain unaware of the stalker, even after its attack.

What animal can successfully attack people, wound them, eat perhaps an ounce of their tissue, and escape undetected?

The leech sensed the girl, tasted and smelled her through its skin. It swam, undulating, to the teenager's bare foot.

With undetectable, gentle contact, the leech fastened its suckers to the delicate skin just below the inner bump of the girl's left ankle. With razor-sharp mouth parts and a local anesthetic, the leech made an incision. Although the girl's brain detected no wound, her blood automatically began clotting. The leech countered by secreting hirudin, an anticoagulant.

The leech would take 20 times its body weight in blood, enough to last it until the following spring, before departing—still undetected!

Only a small, puzzling wound would remain to mark the site of the painless parasite's bite.

Supplemental Activities

1. Respond to this statement: Biting painlessly is a survival skill, not a kindness.
2. Storing and digesting blood over a period as long as six or seven months confronts the leech with some problems. Suggest some of these problems and speculate as to their resolution before researching this topic.
3. Most leeches do not attack humans. Some do not even suck blood. Look into the natural history of these other "bloodsuckers."
4. Find the class name of the leeches and relate it to the anticoagulant the bloodsucking leeches produce.

191. Worm Diggers

The battered old pickup truck rumbled to a stop in the two muddy ruts that passed for a road. The sole occupant, a man in his early 20's, pulled on a pair of hip boots, picked up a bucket and a short-handled, long-tined rake, and walked purposefully across the mud flats to the sea's edge. Without looking left or right, he bent over and began digging. He continued digging just behind the falling tide, and dug until the tide turned several hours later, pushing him back toward his truck.

During the digging, he paused only to pick up worms and deposit them in his bucket.

The worms, some more than a foot long, were shimmery green with bristles clustered on either side of each of the many segments of their bodies.

After passing through several hands, the price increasing with each transaction, these worms would become bait used by surf-fishing people all along the shore of the southern Atlantic.

Supplemental Activities

1. Worm-digging—worm-selling—is a big business for several northeastern coastal states blessed with muddy shores. From conservation departments of northern New England states, obtain up-to-date information on the financial rewards of worm-digging.
2. Investigate the biology of the marine worm.

---

192. The Action of Worms

> The plow is one of the most ancient and most valuable of human inventions; but long before people existed the land was in fact regularly plowed, and still continues to be thus plowed, by earthworms. It may be doubted whether there are many other animals which have played so important a part in the history of the world as these lowly organized creatures.

What famous British scientist of the past century would have uttered these words?

These were the thoughts of a man who had studied earthworms more than any other person of his day, a man whose name will be forever linked with other work: Charles Darwin.

Supplemental Activities

1. What is that "other work" of Charles Darwin upon which his fame rests?
2. Justify Darwin's statement about the plow being one of humans' most ancient and valuable inventions.
3. Report on the results of the action of worms on the soil.
4. Set up an experiment involving soil samples with and without earthworms to test the worms' role as "plows."

Sharks

193. Immune to Cancer?

What are the two scariest words in the English language?

Psychologists tell us that two of the most frightening words in our language are *cancer* and *shark*.

Besides being a pair of scary words, sharks and cancer are linked to a scientific puzzle. Sharks are reportedly immune to cancer. If so, they would be one of only a few kinds of organisms to escape this disease.

Within the blood of sharks are antibodies protecting sharks from chemicals, bacteria, and viruses—including ones unlikely to be encountered by sharks.

Most interesting of the sharks' immunities is their resistance to cancer. Medical research is now being conducted to determine how the scariest fish might help us attack the scariest disease.

Supplemental Activities

1. Find out how solid the research suggesting sharks do not get cancer is. Sometimes, especially with deadly diseases, word of possible cures is not really scientific.
2. With a little research, you may find other things that make sharks worthy of study by medical science.
3. Find out about medical research done with leeches, pitcher plants, lichens, and bees.
4. In terms of this opener, make a statement about the importance of protecting all species from extinction.

Soil

194. More Than Dirt

Quickly list all the things you can think of that are found in soil.

Proportionally, there is more dust on most classroom globes than soil upon the planet. Yet upon that film of soil, much, if not all, the life on the planet depends.

Soil is a boundary between living and nonliving things of the earth: organic and inorganic.

The soil is a complex amalgam of rock particles, from dust to boulders; water; air; decaying organic matter; and living things from bacteria through burrowing mammals. Soil certainly is more than dirt.

Supplemental Activities

1. After studying the variety and roles of soil organisms, expand and clarify this statement: The soil's inorganic materials, essential to life, enter living systems through living systems.
2. Obtain a soil map of your county and collect samples of local soil groups.
3. Collect samples of living things from local soil.

Sponges

195. Sacrificing Rocks to Sea Gods?

Athletically built young Greeks were in each of the boats, some 2,000 years ago. One of the young men in each boat removed all his clothes, then dove or jumped overboard. A remaining man gave each swimmer a huge rock. Immediately, the swimmer and the rock disappeared. Later, long minutes later, only the youths would surface.

Were the boys sacrificing rocks to sea gods? What were they doing?

The young men were using the rocks to carry them to the bottom, where they would search for sponges before their breath gave out.

Often the rocks would be retrieved by ropes so the same weight could be used repeatedly.

In this strenuous, dangerous work, young men aged quickly. Some never had a chance to grow old.

To the Greeks then and to many people now, sponges were worth the risks.

Supplemental Activities

1. What were sponges used for then? What are natural sponges used for now?
2. Consult books on diving and scuba to learn the physical dangers of such work.
3. How do artificial, synthetic sponges compare with natural sponges? in price? capacity? chemical resistance? texture? sales?
4. Find out how sponges are prepared for market.
5. In Florida especially, sponge diving is an ethnic trade. Look into the nationality and traditions of Florida's sponge divers.
6. Obtain specimens of natural and synthetic sponges and conduct your own quality and usefulness tests.

196. Population 17,128

What role do living sponges play in the ecology of the coral reef?

A sponge is a labyrinth of passages and openings of various sizes. Sponges have few natural enemies due to their odor and indigestible composition. Sponges are little threat to any creatures above the size of their planktonic food. For these reasons, sponges are natural dwelling places for many small animals.

A. S. Pearse wondered just how important sponges were to smaller reef creatures. He secured a sponge big enough to fill your kitchen sink, then set about to classify and count every organism in that sponge big enough to see without a microscope.

He counted 17,128 animals! Most were a single kind of shrimp especially adapted to hide and live within the sponge, but many species from most of the phyla were also represented, including several slender, eel-like fishes ranging to five inches in length.

Supplemental Activities

1. What percent of your hometown's population is 17,128?
2. How many towns within a day's drive of your school have a population of 17,128 or less?
3. If possible, take some samples of wildlife and count a population as Pearse did. You could select a rotting log, kelp holdfasts, a bucket of swamp water, a small tide pool, a soil sample, or a square meter of some convenient area such as a bayou or backyard.

197. Boring Sponges

How could a sponge attack an animal larger than itself?

Cliona appears as bright yellow, spongy bumps on the surface of shells washed ashore along the Carolina coasts.

A living oyster covered with these sponges is a doomed mollusk. The larval sponges attach themselves to mollusk shells and begin boring into the shell, probably by chemical means. Tiny pits become passages, and passages become mazes filled with living sponge tissue. The surface bumps are merely openings in the sponge for exchanging water and other materials with the sponge's environment.

Eventually the sponge can so riddle and weaken the shell that it will crumble, killing the mollusk.

Supplemental Activities

1. How might such behavior benefit the sponge?
2. Collect samples of *Cliona* in situ and see for yourself the effects of a boring sponge. Many biological supply houses offer these sponges if a trip to the Carolinas is not convenient.
3. *Cliona* is very important to the ecology of reefs because of this boring. Can you justify that statement?

198. "Beam Me Back, Scotty!"

What do you suppose would happen if we separated every cell of a living sponge from every other cell?

As every Trekkie knows, the *Enterprise*'s transporter disassembles the molecules of its passengers or cargo, sends them to a designated place, then reassembles them.

The idea of an organism being able to withstand cell-by-cell separation may seem like the wildest science fiction. However, for sponges, it is a demonstrable fact.

In a classic experiment, H.V. Wilson forced a sponge through a silk screen, separating the cells from each other. After a brief rest in calm water, the sponge rebuilt itself.

Supplemental Activities

1. What aspect of sponge anatomy makes this sort of thing easier for sponges than it might be for higher organisms?
2. If you have either saltwater or freshwater sponges in an aquarium, attempt to replicate this experiment.
3. Relate this experiment to the various ways sponges reproduce.
4. Look up sponge farming and relate its techniques to sponge anatomy and to Wilson's experiment.

Symbiosis

199. Together We Stand

All around the two wounded men, the battle for Gettysburg raged. A single shellburst had ended their war, blinding the solider in blue, crippling the man in grey. They lay sprawled, talking to each other to ease their pain and fear of dying; hesitantly at first, then increasingly earnestly.

It occurred to them that thirst and exposure to more fire were their real enemies. If they could get to Spangler's Spring at the base of nearby Culp's Hill, they would have water and shelter.

The blind man supported the cripple; the cripple directed the blind man. Together they had the legs and eyes it would take to survive in a hostile world neither could survive in independently. Together they limped to water, safety, and life.

Such partnerships seem very human, but biology is filled with unions of organisms even more different than the soldiers in this story.

What do we call biological partnerships between organisms of different species?

Such arrangements which have evolved between varied organisms are called symbiosis.

Supplemental Activities

1. Find definitions and examples of commensalism, mutualism, and parasitism; three categories of symbiosis.
2. Look up each of these symbiotic pairs and decide what kind of symbiosis they represent: shark and remora; termite and bacteria; giant clam and algae; plume moth and sundew.
3. Why are such close relationships as moss sporophyte and gametophyte *not* considered examples of symbiosis?
4. Examples of symbiosis exist near you, perhaps even within your school. Find one or more symbiotic pairs and rear them in the laboratory for close observation.
5. Pick a symbiosis set of particular interest to you and try to describe the evolutionary steps that would have brought those partners together in symbiosis.

Virus

200. Sick Wolf/Werewolf

How might a viral disease have spawned the werewolf legend?

At first the villagers thought he was lucky to have survived. Few men attacked by the large European wolf were so lucky. But this hardy peasant had clubbed the wolf to death, bound up the wounds on his arms, and walked the moonlit path home.

But by the next full moon, the villagers knew how unlucky their neighbor and they had been.

They had chained their former friend, now a snarling, slobbering wild beast, to protect his family and themselves from the consequences of his terrifying transformation.

What disease had been transmitted from wolf to man?

Today we know the poor man was dying of rabies, but in medieval Europe, among people who had no science to explain such things, the man was a werewolf's victim, becoming a werewolf himself in the eerie light of the full moon.

Supplemental Activities

1. Such a vignette is supposition rather than history. It does, however, offer an explanation of the werewolf myth. Evaluate the story after reading about the etiology, incubation period, and symptoms of rabies.
2. Research other "monster" stories about vampires, changelings, and giants to see if the tales might have scientific explanations.
3. Look up the derivation of the word *rabies* and its popular synonym, *hydrophobia.* Then explain the appropriateness of these names for this disease.
4. Report on the treatment for rabies and ways people can protect themselves against rabies.

Index

Note: References are to page numbers, not opener numbers.

E

F

R

S